From Visual Surveillance
to Internet of Things

From Visual Surveillance to Internet of Things

Technology and Applications

Edited by
Lavanya Sharma and Pradeep K. Garg

CRC Press
Taylor & Francis Group
Boca Raton London New York

CRC Press is an imprint of the
Taylor & Francis Group, an **informa** business

A CHAPMAN & HALL BOOK

CRC Press
Taylor & Francis Group
52 Vanderbilt Avenue
New York, NY 10017

Visit the Taylor & Francis Web site at
http://www.taylorandfrancis.com

and the CRC Press Web site at
http://www.crcpress.com

Dedicated to my Dada Ji

(Late. ShriRam Krishan ChoudharyJi)

Ek prerna mayeh Vyaktitavh

Dr. Lavanya Sharma

Dedicated to my Parents

Prof. Pradeep K. Garg

Contents

Part 3   IoT with Visual Surveillance for Real-Time Applications

Part 4   Challenging Issues

Preface

This book explores the utilization of the Internet of Things (IoT) with visual surveillance and its underlying technologies in different application areas. Using a series of present and future applications—business insights, indoor-outdoor security, smart grids, human detection and tracking, intelligent traffic monitoring, e-health, and many more—this publication encourages readers to gain deeper knowledge about implementing the IoT with visual surveillance.

This book comprises four parts that provide an overview of basic concepts, from the rising prevalence of machines and communication to the IoT with visual surveillance, critical application domains, tools, technologies, and solutions to handle relevant challenges. Detailed topics for readers include practical ideas for using IoT with visual surveillance (motion-based object data) to deal with human dynamics, challenges involved in surpassing diversified architecture, communications, integrity, and security aspects. IoT in combination with visual surveillance has proven to be most advantageous for companies to efficiently monitor and control their day-to-day processes, such as production, transportation, maintenance, implementation, and product distribution.

Overall, this publication, *From Visual Surveillance to Internet of Things: Technology and Applications*, helps readers to understand IoT with visual surveillance and to appreciate its value to individuals as well as organizations.

Acknowledgments

I am especially grateful to *my dada ji, my parents, my husband,* and *my beautiful family* for their continuous support and blessings. Very special thanks to *my sweet cute Romeo Sharma and Vedant* for being a part of my life. I owe my special thanks to **Samta Choudhary ji** and **Pradeep Choudhary ji** for their invaluable contributions, cooperation, and discussions.

Very special thanks to **Shri Parvesh Sahib Singh Verma ji** (**Hon'ble Member of Parliament, India**) for his blessings and invaluable support. He simply gives only direction with hint and it inspired me to explore new ideas.

I am very much obliged to **Prof. P.K. Garg**, second editor of this book, for his motivation and support. This would not have been possible without his blessings and valuable guidance.

Above all, I express my heartiest thanks to *God* (The One To Whom We Owe Everything) **Sai Baba of Shirdi** for all blessings, guidance, and help. I would like to thank God for believing in me and being my defender. Thank you, Almighty God.

Lavanya Sharma

I would like to thank my family members: my wife Mrs. Seema, Dr Anurag, Dr Garima, Mr Hansraj, Ms Pooja, and Master Avyukt for providing their fullest support and cooperation during preparation of this book.

Pradeep K. Garg

About the Editors

Dr. Lavanya Sharma received her M.Tech (Computer Science and Engineering) in 2013 from Manav Rachna College of Engineering, affiliated with Maharshi Dayanand University, Haryana, India. She received her Ph.D. from Uttarakhand Technical University, India, as a full-time Ph.D. scholar in the field of digital image processing and computer vision in April 2018, and also received a TEQIP scholarship for the same. Her research work is on motion-based object detection using background subtraction technique for smart video surveillance. She received several prestigious awards during her academic career.

She has more than 20 research papers to her credit, including Elsevier (SCI Indexed), Inderscience, IGI Global, IEEE Explore, and many more. She has authored a book *Object Detection with Background Subtraction*.

She contributed as an Organizing Committee member and session chair at Springer's ICACDS conferences 2016, ICRAMSTEL 2019, and Springer's ICACDS 2019. She is an editorial member/reviewer of various journals of repute, including Inderscience, IGI Global, and many others. She is an active program committee member of various IEEE and Springer conferences also.

Her primary research interests are digital image processing and computer vision, vehicular ad hoc networks, mobile ad hoc networks, and Internet of Things. Her vision is to promote teaching and research, providing a highly competitive and productive environment in academic and research areas with tremendous growing opportunities for the society and her country.

Professor Pradeep K. Garg worked as a Vice Chancellor, Uttarakhand Technical University, Dehradun (2015–2018). Presently he is working in the Department of Civil Engineering, IIT Roorkee as a professor. He has completed a B.Tech (Civil Engineering) in 1980 and M.Tech (Civil Engineering) in 1982 both from the University of Roorkee (now IIT Roorkee). He is a recipient of the Gold Medal at IIT Roorkee to stand first during M.Tech program, the Commonwealth Scholarship Award for doing Ph.D. from University of Bristol (UK), and the Commonwealth Fellowship Award to carry out post-doctoral research work at the University of Reading (UK). He joined the Department of Civil Engineering at IIT Roorkee in 1982, and gradually advancing his career, rose to the position of Head of the Department in 2015 at IIT Roorkee.

Professor Garg has published more than 310 technical papers in national and international conferences and journals. He has undertaken 26 research projects and provided technical services to 83 consultancy projects on various aspects of Civil Engineering, generating funds for the Institute. He has authored a textbook on *Remote Sensing* and another one on *Principles and Theory of Geoinformatics*, as well as produced two technical films on story mapping. He has developed several new courses and practical exercises in geomatics engineering. Besides supervising a large number of undergraduate projects, he has guided about 70 M.Tech and 26 Ph.D. theses. He is instrumental in prestigious Ministry of Human Resource Development (MHRD)-funded projects on e-learning, Development of Virtual Labs, Pedagogy, and courses under the National Programme on Technology Enhanced Learning (NPTEL). He has served as an expert on various national committees, including Ministry of Environment and Forests, National Board of Accreditation (All India Council of Technical Education), and Project Evaluation Committee, Department of Science and Technology, New Delhi.

Professor Garg has reviewed a large number of papers for national and international journals. Considering the need to train the human resources in the country, he has successfully organized 40 programs in advanced areas of surveying, photogrammetry, remote sensing, geographic information system (GIS), and global positioning system (GPS). He has successfully organized 10 conferences and workshops. He is a life member of 24 professional societies, out of which he is a Fellow member of 8 societies. For academic work, Professor Garg has travelled widely, nationally and internationally.

Contributors

Himanshu Kumar Agrawal
Startup Fellow
Ministry of Human Resource Development
    (MHRD) Innovation Cell, AICTE
New Delhi, India

Naman Kumar Agrawal
Young Professional
Atal Innovation Mission, NITI Ayog
New Delhi, India

Pallavi H. Bhimte
Dept. of CSE
Galgotias University
Greater Noida, India

Thierry Bouwmans
Laboratoire MIA
University of La Rochelle
La Rochelle, France

Belmar García-García
Instituto Politécnico Nacional
Mexico City, Mexico

Pallavi Goel
Dept. of CSE
Galgotias University
Greater Noida, India

Shailja Gupta
Department of Computer Science
    Technology
Manav Rachna University
Faridabad, India

Umang Kant
Krishna Engineering College
Ghaziabad, Uttar Pradesh, India

Sunil K. Khatri
Amity Institute of Information Technology
Amity University
Noida, Uttar Pradesh, India

Dharmendra Kumar
Delhi Technical Campus
Greater Noida, India

Saket Kumar
Amity University
Noida, Uttar Pradesh, India

Sugandhi Midha
Department of Computer Science
    and Engineering
Chandigarh University
Gharuan, Mohali, India

Shailendra Mishra
Majmaah University
Majmaah, Kingdom of Saudi Arabia

Neetu Mittal
Amity Institute of Information
    Technology
Amity University
Noida, Uttar Pradesh, India

S. Ranjan
Amity University
Noida, Uttar Pradesh, India

Riya Sapra
Department of Computer Science
    Technology
Manav Rachna University
Faridabad, India

Aamna Shahab
Amity Institute of Information
    Technology
Amity University
Noida, Uttar Pradesh, India

Bhupesh Kumar Singh
Amity University
Noida, Uttar Pradesh, India

Mayank Singh
University of KwaZulu-Natal
Durban, South Africa

Viranjay M. Srivastava
University of KwaZulu-Natal
Durban, South Africa

Vijay Kumar Tayal
Amity University
Noida, Uttar Pradesh, India

Dileep Kumar Yadav
Dept. of CSE
Galgotias University
Greater Noida, India

Part 1

Introduction to Visual Surveillance and Internet of Things

1

The Rise of the Visual Surveillance to Internet of Things

Lavanya Sharma

Amity Institute of Information Technology,
Amity University, Noida, Uttar Pradesh, India

CONTENTS

1.1 The Rise of the Visual IoT

Today's developments in networking technologies enable usage of networking in motion-based objects in the course of IoT, which results in improvement of living standard. Using these technologies, an object's behavior is automatically detected and an intelligent surveillance system raises an alarm in case of any suspicious, threatening, or illegal activities [1–12]. Surveillance provides secure monitoring using intelligent visual surveillance of human activities, forestry, natural environments, human-machine interactions (HMI), content-based video coding, and many more areas. Intelligent surveillance systems require the usage of both control system and IoT technologies in order to reduce the need for human beings in authorized domains. In this way, users can monitor their particular domains from anywhere with the help of Internet facilities and mobile sensor-based devices. Users have control to operate the intelligent surveillance system from anywhere. Modern closed-circuit television (CCTV) technology uses large amounts of disk storage to reserve video streaming data for future reference [13–18]. In literature, some authors developed and proposed an idea to overcome the storage issue and to send the alert message to the concerned person during any intrusion in that particular authorized domain.

Video-streamed data is a rapid increasingly sphere of big data or cloud computing along with IoT projects, and its importance is rising to match its profusion (see Figure 1.1). Organizations are at a crossroads where many types of new technologies converge to help individuals or concerned persons to utilize the video data that is already collected as well as data collected from new sources. The IoT, cloud computing technologies, big data, and data analytics are the drivers of this new technological world [19–25]. These technologies aid smart cities, driverless cars, and emergency departments. But until recently, few people have been familiar with the key role of visual or video data that is collected from various sources including CCTV footage, sensor-based devices within the IoT. For example, at airport check-ins, a reader scans the boarding pass of a passenger and a camera scans the face of that person using various face detection and recognition techniques. Without the video data that the camera captures, the system would be less useful.

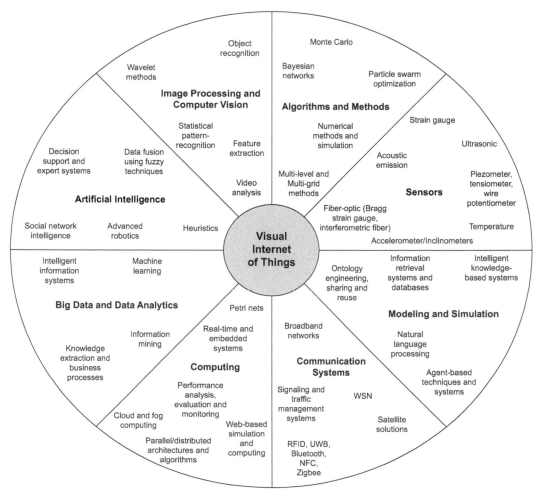

FIGURE 1.1
Enabling networking technologies for the IoT.

1.2 The Rise of Visual Analytics

The increasing demand for video data led to the rise of visual analytics. According to a report by McKinsey and company, visual analytics applications have seen enormous growth: more than 50% over the upcoming 5 years, which will contribute to a prospective economic impact of about $3.9 trillion to $11.1 trillion annually by the end of year 2025 for the IoT. When considering visual analytics, people initially think about visual surveillance [2, 8, 12, 26–29]. Indeed, the skill of visual analytics is to detect any kind of suspicious, threatening, or illegal activity, and also detection of faces; there are ample instances in which intelligent surveillance systems can be advantageous. Recently, cloud and big data technologies drive the world's economy of scale, making such intelligent systems accessible to a great diversity of settings, from the millions of footfalls of railways, airports, rural, and urban zones to smaller locations such as holy places, schools, colleges, emergency departments or medical care centers, hotels, and care homes [16, 20, 22, 30–32].

But visual analytics goes beyond visual surveillance. Visual analytics provides a system to automatically detect and track any suspicious activity. Due to modern advancements in camera technologies, current processing and algorithm development systems allow the software to recognize the most informative objects and other related background details. They also enable analytics to successfully detect moving objects in the video stream except in cases of major challenging issues (e.g., illumination variations, water rippling, sleepy object, slow leafy movements, moved object, dynamic or complex background). Any system using visual analytics can efficiently detect and track motion-based objects in case of emergency and quickly deploy staff to handle the situation [21, 22, 29, 33]. Settings such as smart homes, smart cities, and smart retail can be equipped with smart visual analytics technologies for motion-object detection (human, vehicles, animals), face detection, zonal intrusion, automatic number plate recognition (ANPR), heat maps, dwell analysis, people counting, face recognition and management dashboards, and auditory analysis as shown in Figure 1.2.

1.3 Transitioning Away from Conventional Visual Surveillance

Today, the majority of visual systems still rely on a human operator to monitor video streams and manually detect illegal or suspicious activities. However, in metropolitan

(a)   (b)   (c)

FIGURE 1.2
Artificial intelligence (AI)-based video analytics solutions for smart cameras: (a) smart universities, (b) smart halls, and (c) smart homes.

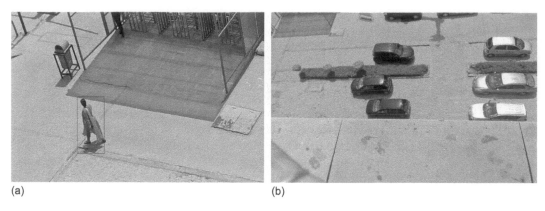

(a)   (b)

FIGURE 1.3
Real-time drone surveillance system: (a) identifying moving persons from a video stream, (b) aerial surveillance for smart cities.

cities, this approach becomes more expensive, untenable, and prone to much human error. Conventional technology such as traditional video motion detection (VMD) detects changes in backgrounds and alerts security officers of suspicious activities. At times, natural elements such as dynamic backgrounds, slow leafy movements, and cloudy environments change the video stream's pixel values, which results in false alarms. If these data were sent to the cloud and visual analytics were applied to track breaches, the number of false alarms would decrease. Moreover, this conventional technology cannot be used for large amounts of data, especially in metropolitan cities where the number of objects is higher. The potential for the visual IoT is enormous. Examples include unmanned aerial vehicles (UAVs), or drones, equipped with cameras that can help disaster-relief firms discover available routes through destroyed streets; identifying vicious persons in public domains as shown in Figure 1.3; and classifying new scenes. Outdoor or indoor cameras can efficiently monitor object movements not to "spy" on people, but to ensure they are in motion, and no suspicious activity is occurring [4, 5].

According to a report, the UK's environment secretary said that CCTV will become mandatory in all abattoirs in England. The goal of this decision is to increase animal-welfare standards, and video data can be easily accessible to the food standards firms and veterinarians, who will monitor and enforce animal-welfare standards in the abattoirs. The government also confirmed that this will result in improved standards for farm animals and domestic pets by streamlining statutory animal-welfare codes to reflect improvements in medicine, technology, research, and guidance from veterinarians [3].

1.4 A Watershed for Legacy Systems

Per the report of Cloudview, currently 8.5 million visual cameras in the UK are used for different surveillances. By combining all, nearly 10.3 petabytes of visual data are created every single hour; mostly, these collected data are stored locally and never even viewed. Some, but not all, of these cameras are found to be appropriate for visual analytics. About 2 million transportation systems in the UK can track moving persons, medical

(a)   (b)

FIGURE 1.4
Internet protocol-based commercial visual surveillance for security: (a) indoor environment, (b) outdoor environment.

centers, and roadside safety, as they are for conventional CCTV technique use only. About 800,000 video cameras in the housing zone currently store their visual data on a local storage device, where it has to be accessed manually. Thus, a person must be present to check the records of current and historical data [6].

Those housing providers who store their visual data in the cloud unlock innovative potential such as using visual analytics to automatically check the location's day-to-day activities including both indoor and outdoor, maintenance, and parking allotments. Concerned staff members or security officers can be alerted as necessary in real-time, making it much easier for them to address the issues. In the case of serious problems, they can also share video sequences with agencies such as ambulance, police, and fire services [11, 14, 34] as shown in Figure 1.4.

1.5 Tools and Technologies

Society is at a turning point where numerous types of technologies are converging, which will help the community to make more utilization of the visual data that are collected from various sources. From the food people eat, transport used for daily routines, activities in shopping complexes and homes, visual data can play an important part. Some of the important technologies are discussed in this section.

1.5.1 IoT and Computer Vision

Computer vision technologies are widely used today in real-time scenarios ranging from game consoles that can efficiently recognize the player's gestures to mobile cameras that can automatically focus on the object [24, 35, 36]. This capability impacts many areas of our lives. In fact, this technology has a long historical background in both the commercial sector and government. Sensor-embedded devices that can easily sense light waves in various spectrum ranges are widely used in various real-time applications such as manufacturing firms for quality assurance, object detection and tracking for security in visual surveillance, and remote sensing for environmental management (drone, satellites, or other vehicles). In the past, many of these applications were limited to only selected platforms. But when combined with Internet protocol (IP)-connected devices, they create a new set of real-time-based applications that were not implemented before [19, 21, 31, 37–39]. Computer vision techniques with IP devices, advanced data analytics, and AI, led to a revolutionary hike in IoT innovations and applications.

Working in commercial surroundings has always been unsafe. Security personnel have long been looking for solutions to minimize the risk of casualties. Their motive is to reduce vulnerability, handle risks, and avert accidents. Using sensor-based devices it becomes easier to monitor real-time scenarios (i.e., in case of any suspicious activity in industrial or other restricted places, appropriate timely decisions can be made) [17, 18, 32, 40]. For example, if sensors detect leakage of gaseous elements, increased temperatures, or surplus humidity, work can be halted at once or at the very least the security authorities can be updated. These type of decisions are deterministic and do not provide much insight into the future. Another way of creating a safer environment is to use machine-learning technologies with network technologies. By creating different situations, the algorithm or technique can sense the difference between safe and dangerous scenarios. With the advancements in computer vision, machine learning, deep learning, and image processing technologies, the proposed algorithms or methods can efficiently detect motion-based objects and velocity in real-time applications [21, 22, 41]. With the improvement of processing power (GPUs), sensor-based devices, and independent-carry systems such as robots and drones, we have safety routines that are fully automatic and better than their human counterparts.

1.5.2 IoT and Remote Sensing

Remote sensing is one of the most promising technologies for the IoT, in which several imaginable entities can be easily equipped with unique identifiers and autonomously transfer all the crucial information or data over a network. This technology is a subdomain of geography. Contrary to on-site observation, remote sensing involves acquiring information about a particular object or a phenomenon without any bodily contact with that object [29, 42]. Sensing in terms of this technology refers to the utilization of aerial sensor-based technologies to detect and classify the targeted objects on earth including both the outer and inner surface of earth (atmosphere and oceans) by means of propagated signals such as electromagnetic radiation. This can be further categorized into active remote sensing, where devices are fitted with transmitters that propel signals of a particular wavelength or electrons to be bounced off the targeted object, data are gathered by the sensor upon their reflection and secondly, passive (sunlight) where light emitted from the sun can be captured by sensors such as a charge-coupled device (CCD) camera mounted on a satellite [10].

These technologies mainly include light detection and ranging (LiDAR), radar, infrared (IR) radiation, thermal radiation, sonar, electric field sensing, seismic, and global positioning system (GPS). Depending on the targeted object that is to be detected, these are mounted to a satellite, submarine, boat, UAV drone or airplane, or from another convenient observation point such as the top-most level of a building [19, 20, 32]. The data gathered by these technologies can be used for a large number of real-time applications including exploration of several resources, cartography, measurements of atmospheric chemicals, visual surveillance, navigation, healthcare observation, and GPS tracking [43] as shown in Figure 1.5.

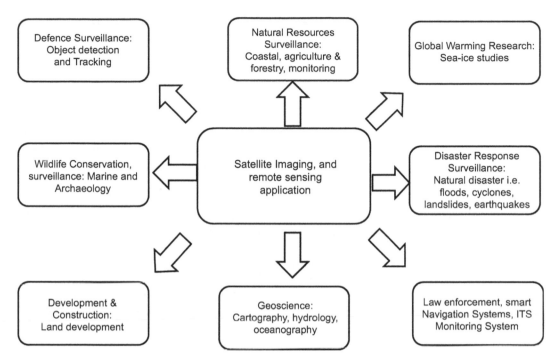

FIGURE 1.5
Various applications of remote-sensing technology.

1.5.3 IoT and Artificial Intelligence

IoT is a most promising technology these days. It is predicted that by the end of 2020, there will be about 30 million IoT gadgets or devices all over the globe. AI is the branch of computer science that deals with some sort of intelligence established by machines and also tries to impersonate the natural intelligence of a person, which means it imitates cognitive functions that human beings perform using their intelligence such as learning, problem solving, and many more based on supervised or unsupervised learning paradigms [35, 36].

Both AI and IoT are some of the trendiest technologies, which are widely used by various companies. AI makes the machines learn from the knowledge stored in their knowledge base stored data based on experiences of knowledge experts, and IoT is about sensor devices interacting with the other devices over IP, and these devices are generating approximately 3.5 quintillion bytes of data every day. These data become a powerhouse of collected facts and information on a daily basis. Further, the data can be used to build predictive models, and conduct experimental analysis such as smart assistants used in smart homes to manage household tasks [15, 44, 45]. For example, if your milk carton is nearly empty in your refrigerator, the smart assistant will "see" this information via sensors and use the Internet to alert you or order more milk. This is a classic real-time example of smart assistants enriching our lives by using both technologies IoT and AI.

Recently IBM launched the Watson Assistant, a smart assistant that combines both IoT and AI technologies to do wonders for smart businesses. It can be accessed by either text or voice, and tries to help business entrepreneurs or employees to increase loyalty to their brands and mend the client experiences while keeping their data secure. The Watson Assistant is pre-trained for various real-time applications such as emergency departments, customer care, banks, and indoor scenes. SoftBank Robotics implemented the Watson Assistant technology for its humanoid robot named "Josie Pepper," which is used at the Munich Airport to assist passengers regarding weather information at their destinations, to make small talk, and to provide other relevant information to travelers [46].

1.5.4 IoT and Cloud Computing

Cloud computing and the IoT share a complementary relationship and serve to boost effectiveness in daily tasks. Each day, IoT generates huge amounts of data and cloud computing provides a lane for these data to travel. Economies of scale are another means where cloud providers can benefit small-scale IoT startups and lessen the all-inclusive outlay to IoT companies. An additional advantage is that cloud computing in combination with IoT enables better collaboration for developers [13, 14, 47, 48]. Because developers can accumulate and access this collected data remotely, they can access this collective data immediately and may work on several associated projects without any delay. Finally by storing this collective data in the cloud, IoT firms can modify directly and speedily and allocate resources to different domains [45].

The growths of both technologies are interconnected, which creates a huge connection of "things." The combination of cloud computing and IoT will facilitate innovative monitoring services and processing of data collected from sensor devices. For example, the collected data from various sensor devices can be uploaded and stored with cloud computing, then these data can be used intelligently for any smart monitoring purpose and alarm alert with other smart gadgets or devices. Ultimately, the objective is to be able to transform digital data to insights, and transform insights to cost-effective deeds [16, 30, 31, 49]. Here, the cloud efficiently serves as the brain in order to improve decision-making power and also to optimize Internet-based interactions. But when IoT is integrated with the cloud, several new challenges arise, such as quality of service (QoS), quality of experience (QoE), data security, reliability, efficiency, and integrity [50].

1.5.5 IoT and Big Data

The term *Internet of Things* refers to an ecosystem of smart gadgets or devices that are interconnected with one another. Basically, it is a network of sensory devices having IP addresses and capable of generating, transmitting, storing, and receiving millions of data daily without any human assistance. The next logical question becomes, Where does all this data get processed then? This is where big data steps in. The term *big data* describes datasets in which large amounts of data that are generated by the IoT can be processed easily and efficiently. IoT generates data in such massive amounts that it qualifies as big data [12, 27, 51]. IoT devices are smart sensory devices and they report reverse to aggregation processes which are basically devices with high processing power and perform tasks of aggregation. Big data is also analyzed in machines in the cloud where their flexibility of accumulating the computing power on demand makes it quite simpler, easier, and more feasible to implement several analytics on sorted data units as shown in Figure 1.6.

1.5.6 IoT and Ubiquitous Computing

In 1988, Mark Weiser introduced ubiquitous computing (UC) as the idea of workstations being everywhere, running invisibly in the background. IoT is the current frontrunner in bringing this idea to life. During the last few epochs, the development of computation tools

FIGURE 1.6
Block diagram of IoT gateway.

and technology produced more powerful storage capacity, processors, and memory at very low cost. Eventually, several physical things will be embedded with additional computation capabilities, and these items can be easily interconnected over a network [41, 52]. This advancement leads IoT to a new level. A lack of standards is easily depicted in ecosystems in both industrial and research zones. The important factors of collaboration and interoperability between various IoT-based applications for real-time systems enrich the concept of UC. Examples on the market now include LIFX Wi-Fi lighting, ConnectSense smart outlets, and Eyedro smart home electricity monitor devices to help homeowners save electricity in an efficient manner [25, 52, 53].

1.6 Conclusion

IoT and visual surveillance both are key technologies that are most commonly used in realistic scenarios for various security aspects. This chapter summarized other promising technologies in combination with IoT and their applications, which results in an improved ecosystem. This chapter provides a detailed overview of how these technologies in combination with IoT modernize the existing scenario to allow detecting and tracking moving objects, alerting to suspicious activities in outdoor and indoor places and restricted zones, and many more strategies to enhance security. Various resources are available for more detailed reading [54–89] on topics covered in this chapter.

References

1. IoT Overview. https://www.mdpi.com/1424-8220/17/6/1457/htm//
2. Use of IoT in Industries. https://www.mckinsey.com/industries/high-tech/our-insights/video-meets-the-internet-of-things
3. Visual Surveillance for Animal Welfare Plan. https://www.gov.uk/government/news/mandatory-cctv-in-all-slaughterhouses-under-new-animal-welfare-plans
4. Eye in the Sky: Real-time Drone Surveillance System (DSS) for Violent Individuals Identification Using ScatterNet Hybrid Deep Learning Network. https://arxiv.org/abs/1806.00746
5. Drone Overview for Visual Surveillance. https://www.businessinsider.com/paul-szoldra-darpa-drone-sees-insane-detail-from-17500-feet-2014-12?IR=T
6. Introduction to Cloud. http://www.cloudview.co/overview

7. Overview of Commercial Video Surveillance. http://www.superioralarm.com/commercial-video-surveillance-system

8. Visual Surveillance with Internet Protocol. https://srpl.in/tag/ip-based-video-surveillance/

9. Real-time Surveillance Scenario. https://www.youtube.com/watch?v=v06-7E35FEA

10. IoT with Remote Sensing Overview. https://www.iotone.com/term/remote-sensing/t630

11. van Kranenburg, R. (2008). *The Internet of things: A critique of ambient technology and the all-seeing network of RFID.* Amsterdam, The Netherlands: Institute of Network Cultures.

12. Koshizuka, N., & Sakamura, K. (2010). Ubiquitous ID: Standards for ubiquitous computing and the Internet of things. *IEEE Pervasive Computing, 9*(4), 98–101.

13. Atzori, L., Iera, A., & Morabito, G. (2016). The Internet of things: A survey. *Computer Networks, 54*(15), 2787–2805.

14. Wu, M., Lu, T. J., Ling, F. Y., Sun, J., & Du, H. Y. (2010). Research on the architecture of Internet of things. *IEEE 2010 3rd International Conference on Advanced Computer Theory and Engineering, 5*, 484–487.

15. Sharma, L., & Lohan, N. (2017). Performance enhancement through handling of false classification in video surveillance. *Journal of Pure and applied Science & Technology, 7*(2), 9–17.

16. Kokkonis, G., Psannis, K. E., Roumeliotis, M., & Schonfeld, D. (2016). Real-time wireless multisensory smart surveillance with 3D-HEVC streams for internet-of-things (IoT). *Journal of Supercomputing, 73*(3).

17. Akula, A., Khanna, N., Ghosh, R., Kumar, S., Das, A., & Sardana, H. K. (2013). Adaptive contour based statistical background subtraction method for moving target detection in infrared video sequences. *Infrared Physics & Technology, 63,* 103–109.

18. Shah, M., Deng, J., & Woodford, B. (2014). Video background modeling: Recent approaches, issues and our solutions. *Machine Vision and Applications, 25*(5), 1105–1119.

19. de Rango, F., Fazio, P., Tropea, M., & Marano, S. (2008). Call admission control for aggregate MPEG-2 traffic over multimedia geo-satellite networks. *IEEE Transactions on Broadcasting, 54*(3), 612–622.

20. Cristani, M., Farenzena, M., Bloisi, D., & Murino, V. (2010). Background subtraction for automated multisensor surveillance: a comprehensive review. *Eurasip Journal on Advances in Signal Processing, 43,* 1–24.

21. Sharma, L., & Yadav, D. (2016). Histogram based adaptive learning rate for background modelling and moving object detection in video surveillance. *International Journal of Telemedicine and Clinical Practices, Inderscience, 2*(1), 74–92.

22. Sharma, L. (2018). *Object detection with background subtraction.* Latvia, European Union: LAP LAMBERT Academic Publishing (ISBN: 978-613-7-34386-9).

23. Sharma, L., & Lohan, N. (2018). Performance analysis of moving object detection using BGS techniques. *International Journal of Spatio-Temporal Data Science, 1*(1), 74–92.

24. Yadav, D. K., Sharma, L., & Bharti, S. K. (2014). Moving object detection in real-time visual surveillance using background subtraction technique. In IEEE, 14th International Conference in Hybrid Intelligent Computing (HIS-2014). Kuwait: Gulf University for Science and Technology.

25. Carvalho, C. M., Rodrigues, C., Armando, P., & de Castro, M. F. (2015). Adaptive tracking model in the framework of medical nursing home using infrared sensors. In 2015 IEEE Globecom Workshops (GC Workshops), pp. 1–6.

26. Report on Spaceborne SAR Tomography: Application in Urban Environment, https://ieeexplore.ieee.org/document/5497277

27. Fuqaha, A. A., et al. (2015). Internet of things: A survey on enabling technologies protocols and applications. *IEEE Communications Surveys & Tutorials, 17*(4), 2347–2376.

28. Lu, Y. X., Chen, T. B., & Meng, Y. (2011). Evaluation guideling system and intelligent evaluation process on the Internet of Things. *American Journal of Engineering and Technology Research, 11*(9), 537–541.

29. Yadav, K. D., Singh, K., & Kumari, S. (2016). Challenging issues of video surveillance system using Internet of things in cloud environment. In International Conference on Advances in Computing and Data Sciences, ICACDS 2016, pp. 471–481.

30. Sharma, S., Chang, V., Tim, U. S., Wong, J., & Gadia, S. (2016). Cloud based emerging services systems. *International Journal of Information Management*, 1–12.
31. Sharma, S. (2016). Expanded cloud plumes hiding big data ecosystem. *Future Generation Computer Systems, 59,* 63–92.
32. Bao, X., Zinger, S., Wijnhoven, R., & de With, P. H. N. (2013). Ship detection in port surveillance based on context and motion saliency analysis. In Proceeding SPIE, Video Surveillance and Transportation Imaging Applications, pp. 1–8.
33. Dong, X., Huang, X., Zheng, Y., Bai, S., & Xu, W. (2014). A novel infrared small moving target detection method based on tracking interest points under complicated background. *Infrared Physics & Technology, 65,* 36–42.
34. Gigli, M., & Koo, S. (2011). Internet of things: Services and applications categorization. *Advances in Internet of Things, 1*(2), 27–31.
35. Sharma, L., Yadav, D. K., & Bharti, S. K. (2015, February 19-20). An improved method for visual surveillance using background subtraction technique. In IEEE, 2nd International Conference on Signal Processing and Integrated Networks (SPIN-2015). Noida, India: Amity University.
36. Sharma, L., Lohan, N., & Yadav, D. K. (2017, June 5). A study of challenging issues on video surveillance system for object detection. In International Conference on Electrical, Electronic Communication, Industrial Engineering and Technology Management Collaboration: Breaking the Barriers, New Delhi: JNU.
37. Roy, S., Bose, R., & Sarddar, D. (2015). A fog-based DSS model for driving rule violation monitoring framework on the Internet of things. *International Journal of Advanced Science and Technology, 82,* 23–32.
38. Popovic, G., Arsic, N., Jaksic, B., Gara, B., & Petrovic, M. (2013). Overview, characteristics and advantages of IP camera video surveillance systems compared to systems with other kinds of camera. *International Journal of Engineering Science and Innovative Technology, 2*(5), 356–362.
39. Siebel, N. T. (2003). Design and implementation of people tracking algorithms for visual surveillance applications. University of Reading.
40. Li, D., Babcock, J., & Parkhurst, J. (2006). Low cost eye tracking for human computer interaction. http://thirtysixthspan.com/openEyes/MS-Dongheng-Li-2006.pdf
41. Rossana, A., Maia, R., Linhares de Arauja, I., Oliveira, K., & Maia, M. (2017). What changes from ubiquitous computing to Internet of things in interaction evaluation? *Ambient and Pervasive Interactions,* 3–21.
42. Shao, Z., Cai, J., & Wang, Z. (2018). Smart monitoring cameras driven intelligent processing to big surveillance video data. *IEEE Transactions on Big Data, 4*(1), 105–116.
43. IoT with Remote Sensing Overview. https://internetofthingsagenda.techtarget.com/definition/remote-sensing, https://ieeexplore.ieee.org/document/8127491
44. IoT with Artificial Intelligence (future assistance). https://yourstory.com/2018/08/iot-ai-combine-smart-assistants-future/
45. IoT with Cloud Computing. https://pinaclsolutions.com/blog/2017/cloud-computing-and-iot
46. Robotics at Munich Airport. https://www.munich-airport.com/hi-i-m-josie-pepper-3613413
47. Panasetsky, D., Tomin, N., Voropai, N., Kurbatsky, V., Zhukov, A., & Sidorov, D. (2015, June 29-July 2). Development of software for modelling decentralized intelligent systems for security monitoring and control in power systems. In PowerTech 2015 IEEE Eindhoven, pp. 1–6.
48. Scardapane, S., Scarpiniti, M., Bucciarelli, M., Colone, F., Mansueto M. V., & Parisi, R. (2015). Microphone array based classification for security monitoring in unstructured environments. *International Journal of Electronics and Communications, 69*(11), 1715–1723.
49. de Rango, F., & Veltri, F. (2009). Two-level trajectory-based routing protocol for vehicular ad hoc networks in freeway and Manhattan environments. *Journal of Networks, 4*(9), 866–880.
50. Real-time Examples of Cloud Computing with IoT. https://www.newgenapps.com/blog/top-10-cloud-computing-examples-and-uses
51. IoT Real-time Applications. http://www.rfwireless-world.com/IoT/IoT-Gateway.html

52. Araújo, I. L., Santos, I. S., Filho, J. B. F., Andrade, R. M. C., Neto, P. S. (2017). Generating test cases and procedures from use cases in dynamic software product lines. In Proceedings of the 32nd ACM SIGApp Symposium on Applied Computing.

53. Dong, T., Churchill, E. F., & Nichols, J. (2016). Understanding the challenges of designing and developing multi-device experiences. In Proceedings of the 2016 ACM Conference on Designing Interactive Systems, ACM (2016), pp. 62–72.

54. Xiaojiang, X., Jianli, W., & Mingdong, L. (2015). Services and key technologies of the internet of things. *ZTE Communications, 2,* 11–16.

55. Tan, L., & Wang, N. (2010). Future Internet: The Internet of things. *IEEE 2010 3rd International Conference on Advanced Computer Theory and Engineering, 5,* 376–380.

56. Want, R. (2004). Enabling ubiquitous sensing with RFID. *Computer, 37*(4), 84–86.

57. Welbourne, E., et al. (2009). Building the Internet of things using RFID: The RFID ecosystem experience. *IEEE Internet Computing, 13*(3), 48–55.

58. Jia, X. L., Feng Q. Y., Ma, C. Z. (2010). An efficient anti-collision protocol for RFID tag identification. *IEEE Communications Letters, 14*(11), 1014–1016.

59. Culler, D. (2003). 10 emerging technologies that will change the world. *MIT Technology Review,* 33–49.

60. Bang, O., Choi, J. H., & Lee. D. (2009). Efficient Novel Anti-collision Protocols for Passive RFID Tags. Auto-ID Labs White Paper WPHARDWARE-050, MIT.

61. Banks, J., Pachano, M., & Thompson, L. (2007). *RFID applied.* New York: John Wiley & Sons, Inc.

62. Pateriya, R. K., & Sharma, S. (2011). The evolution of RFID security and privacy: A research survey. In International Conference on Communication Systems and Network Technologies (CSNT), pp. 115–119.

63. Bahga, A., & Madisetti, V. (2015). *Internet of things: A hands-on approach.* Universities Press.

64. Bhaskar, P., & Uma, S. K. (2015). Raspberry Pi home automation with wireless sensors using smart phones. *IJCSMC, 4*(5), ISSN 2320-088X.

65. Imran Quadri, S. A., & Sathish, P. (2017). IoT based home automation and surveillance system. In International Conference on Intelligent Computing and Control Systems (ICICCS).

66. Stergiou, C., Psannis, K. E., Plageras, A., & Kokkonis, G. (2017). Architecture for security monitoring in IoT environments. In IEEE 26th International Symposium on Industrial Electronics (ISIE).

67. Lee, U., Zhou, B., Gerla, M., Magistretti, E., Bellavista, P., & Corradi, A. (2006). MobEyes: Smart mobs for urban monitoring with a vehicular sensor network. In IEEE Wireless Communications, pp. 52–57.

68. Batalla, J. M., Krawiec, P., Beben, A., Wisniewsi, P., & Chydzinski, A. (2016). Adaptive video streaming: Rate and buffer on the track of minimum rebuffering. *IEEE Journal on Selected Areas in Communications, 34*(8), 2154–2167.

69. Stergiou, C., Psannis, K. E., Kim, B-G., & Gupta, B. (2016, December). Secure integration of IoT and cloud computing. *Future Generation Computer Systems.*

70. Batalla, J. M., & Krawiec, P. (2013). Conception of ID layer performance at the network level for Internet of Things, *Pers Ubiquitous Computing, 18,* 465–480.

71. Kokkonis, G., Psannis, K. E., Roumeliotis, M., & Ishibashi, Y. (2015). Efficient algorithm for transferring a real-time HEVC stream with haptic data through the internet. *Journal of Real-Time Image Processing, 12*(2).

72. Cruz, T., Barrigas, J., Proenca, J., Graziano, A., Panzieri, S., Lev, L., & Simoes, P. (2015). Improving network security monitoring for industrial control systems. IEEE International Symposium.

73. Mateos, A. C., & González, C. M. (2016). Physiological response and sulfur accumulation in the biomonitor Ramalina celastri in relation to the concentrations of SO_2 and NO_2 in urban environments. *Microchemical Journal, 126,* 116–123.

74. Richards, J., Reif, R., Luo, Y., & Gan, J. (2016). Distribution of pesticides in dust particles in urban environments. *Environmental Pollution, 214,* 290–298.

75. Hofman, J., et al. (2016). Ultrafine particles in four European urban environments: Results from a new continuous long-term monitoring network. *Atmospheric Environment, 136,* 68–81.

76. Santamaria, A. F., Raimondo, P., & Palmieri, N., Tropea, M., & De Rango, F. (2019). Cooperative video-surveillance framework in Internet of things (IoT) domain. In *The Internet of Things for Smart Urban Ecosystems* (p. 305–331). Springer, Cham.

77. Cicirelli, F., Guerrieri, A., & Spezzano, G. (2017). An edge-based platform for dynamic smart city applications. *Future Generation Computer Systems, 76,* 106–118.

78. Song, B., Hassan, M. M., Tian, Y., Hossain, M. S., & Alamri, A. (2015). Remote display solution for video surveillance in multimedia cloud. In *Multimedia Tools and Applications,* pp. 1–22.

79. Gubbi, J., Buyya, R., Marusic, S., & Palaniswami, M. (2013). Internet of things (IoT): a vision, architectural elements, and future directions. *Future Generation Computer Systems, 29*(7), 1645–1660.

80. Ali, S., & Shah, M. (2008). Floor fields for tracking in high density crowd scenes. In Proceedings of ECCV.

81. Buyya, R., Yeo, C.S., Venugopal, S., Broberg, J., & Brandic, I. (2009). Cloud computing and emerging IT platforms: vision, hype, and reality for delivering computing as the 5th utility. *Future Generation Computer Systems, 25*(6), 599–616.

82. Molinaro, A., De Rango, F., Marano, S., & Tropea, M. (2005). A scalable framework for in IP-oriented terrestrial-GEO satellite networks. *IEEE Communications Magazine, 43*(4), 130–137.

83. Gao, G., Wang, X., & Lai, T. (2015). Detection of moving ships based on a combination of magnitude and phase in along-track interferometric SAR-Part I: SIMP metric and its performance. *IEEE Transactions on Geoscience Remote Sensing, 53*(7), 3565–3581.

84. Choksi, M., Zaveri, A.M., & Anand, S. (2013). Traffic surveillance for smart city in Internet of things environment. In Proceedings of SAI Intelligent Systems Conference.

85. IntelliSys. (2018, November). Intelligent Systems and Applications, 189–204.

86. Bramberger, M., Brunner, J., Rinner, B., & Schwabach, H. (2004). Real-time video analysis on an embedded smart camera for traffic surveillance. In Real-Time and Embedded Technology and Applications Symposium: 10th IEEE Proceedings, IEEE (2004), RTAS 2004, pp. 174–181.

87. Ihaddadene, N., & Djeraba, C. (2008). Real-time crowd motion analysis. In 19th International Conference on Pattern Recognition, ICPR 2008, pp. 1–4.

88. Allan, W. (2009). Aerial Surveillance and Fire-Control System 13 Feb 1973, US Patent 3,715,953.

89. Sharma, L., Singh, A., & Yadav, D. K. (2016). Fisher's linear discriminant ratio based threshold for moving human detection in thermal video. *Infrared Physics and Technology, 78,* 118–128.

2

A Novel Internet of Things Access Architecture of Energy-Efficient Solar Battery Charging System for Mobile Phones

Aamna Shahab and Neetu Mittal

Amity Institute of Information Technology,
Amity University, Noida, Uttar Pradesh, India

CONTENTS

2.1 Introduction

The IoT has become vast and proven as a backbone of communication and for industry products. It mainly deals with business analytics, machine learning, and so on, used for applications as well as embedded items. Batteries of mobile devices do not remain charged for a long period of time, and there are times when people urgently require their cell phones but the batteries are not charged. Furthermore, users also may forget to charge their power banks. When people connect a cell phone to a power bank and place it in their pockets, it is inconvenient and cumbersome to carry both devices.

The proposed work discusses the utilization of mobile chargers seamlessly; when users are outdoors they can capture solar power and wireless power transmission to charge devices with ease and comfort. Solar energy has been used to charge devices, as it is a freely available, sustainable source of energy that has diverse applications and low costs of maintenance. The focus is mainly on increasing the mobility of users and making use of solar power as it is a renewable source of energy to save electrical power consumption. To make their devices work, people will have to keep the mobile charger in daylight for a small period and then charge their mobile devices.

A solar panel consisting of monocrystalline silicon cells is synched to a wireless charging pad that uses sensor chips. Afterwards, a cell phone can be revived by simply putting the solar wireless power transmission (SWPT) device on the back of the cell phone with the help of metal plates, which will be configured on both devices. Electron gap sets as soon as light falls on the solar cell. They are made with the n-type emitter and in the p-type base. Battery charging is crucial these days for laptops, cell phones, and many other mobile devices. The energy around us can be used wisely to help people realize the importance of renewable sources, while at the same time making operation less costly as solar power conserves electricity considerably [1]. The proposed work may help people when they are outside their homes, as they can either put the system on the back of their cell phone and keep using it simultaneously or they can keep the device separately in the sun for a few minutes and charge their mobiles later.

The proposed device is accomplished with the help of four main components.

2.1.1 Wireless Power Transmission

Wireless power is truly conduction of electrical energy without wires, and it has been an exciting improvement in consumer electronics, supplanting wired chargers as shown in Figure 2.1. It is a new and innovative field of research. In this chapter a technique to convey energy that is viable and secure to a separation without interruption has been proposed. Usages include long-distance driving vehicles, the transmission of solar power from space and remote battery charging.

Wireless power transfer (WPT) makes it conceivable to supply power through an air hole, without the requirement of current-carrying wires. WPT can give control from an alternating current (AC) source to compatible batteries or gadgets without physical connectors or wires. It can charge various hardware devices. It might even be conceivable

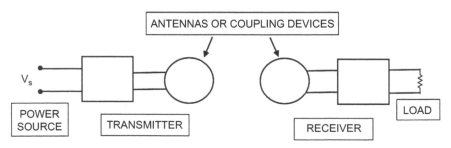

FIGURE 2.1
Wireless power transmission.

to wirelessly transmit power accumulated by solar panels. The electromagnetic fields associated with WPT can be quite strong, and therefore human well-being must be taken into consideration. Contact with electromagnetic radiation can be a great concern, and there is additionally the likelihood that the fields produced by WPT transmitters could meddle with wearable or embedded medical devices. The transmitters and collectors are inserted inside WPT devices, just like the batteries are to be charged. The real charged circuitry will rely upon the technology utilized. Notwithstanding, the framework of genuine exchange of energy must enable the transmitter and recipient to communicate. This guarantees a recipient can inform the charging device when a battery is completely charged. Communication with the charge controller enables a transmitter to distinguish and recognize a receiver, to change the measure of power transmitted to the load, and to monitor, for example, battery temperature.

The idea of near-field versus far-field radiation is pertinent to WPT. Transmission strategies, the measure of intensity that can be exchanged and proximity, are affected by whether the framework is using near-field or far-field radiations. A power-hungry battery significantly limits the lifetime of IoT devices [2]. The energy in the close field is non-radiative, and the oscillating magnetic and electric fields are autonomous to each other. Capacitive (electric) and inductive (magnetic) coupling can be utilized to exchange capacity to a receiver situated in the transmitter's close field. Regions with greater than about two wavelengths of separation from the antenna are in the far field. The best examples are systems that utilize high-control lasers or microwave radiation to exchange vitality over long separations. The focus is to produce scheduling mobile chargers for rechargeable sensor nodes [3] in an efficient way.

The fundamentals of WPT include inductive energy transmission. Copper cable hoists generate the magnetic field by alternating current in the transmitter section. The field can prompt an alternating current in it when a successor coil is put inside the nearby vicinity of that field. The prototype of green electrical resources [4] has been developed to enhance it using IoT [4]. A real-time solar-array monitoring model based on IoT connectivity and cloud computing is proposed [5]. WPT techniques are discussed in the next sections.

2.1.1.1 Microwave Power Transmission

The concept of WPT is shown in Figure 2.2, and given by William C. Brown [6].

WPT contains two areas: the sending phase and the receiving phase. Within the transmission space, the microwave power supply produces microwave power being controlled by an electronic circuit. The function of the waveguide circulator is to protect the microwave supply from the reverse power through a coaxial waveguide adapter. The function of the tuner is to maintain the impedance between the microwave power source and transmitting antenna. At the same time, the directional coupler isolates the attenuated signals. The power is transmitted through the transmitting antenna to the receiving antenna. The receiver section transforms the microwave power into direct current power. The channel resistance coordinating circuit is accommodated by setting the yield ohmic resistance of a signal supply that is reminiscent of an amending circuit.

2.1.1.2 Inductive Coupling

An inductive coupling strategy is the most critical technique for wireless exchange of energy. The power is transferred by means of a transformer based on induction.

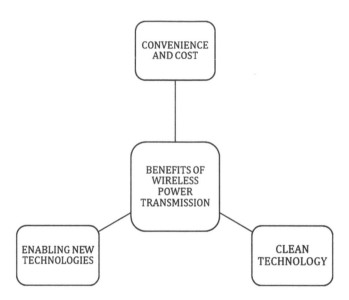

FIGURE 2.2
Benefits of wireless power transmission.

2.1.1.3 Laser Power Transmission

In this type of intensity transmission technique, an optical maser is employed to maneuver light energy; therefore, the power is modified to electrical energy at the recipient finish. The scale and state of the beam are chosen by an arrangement of optics. The transmitted optical maser light is received by the photovoltaic cells, which convert the light into electrical signals. Usually, it utilizes fiber optic wires for transmission.

2.1.2 Photovoltaic Solar Panel

Photovoltaic (PV) solar panels assimilate daylight as a root of energy to produce power. A PV module is a bundled and correlated collection of typically 6 × 10 PV solar cells. Hence, energy produced from the PV model will move towards the wireless charging device through WPT. This energy will then reach out to the battery of the device that requires a charge. In a PV cell, the base and emitter are usually p-type and n-type, respectively. The electron hole pairs are created within the base and emitter. These holes and electrons diffuse at the junction and are moved away by an electric potential. Thus, electric power is transferred.

Solar energy is the next-generation power interface framework, and solar PV panels have an exceptionally encouraging future both for efficient suitability and ecological maintainability. Because solar energy meets the energy requirements for cooling, PV panels can provide answers for energy requirements, particularly in sweltering summer months where energy demands are high.

2.1.3 Solar Power

Solar power converts daylight energy into power. A concentrated solar-based framework utilizes the focal points or mirrors and follows the structures to center a huge region of daylight into a tiny shaft. PV cells convert light into an electric current utilizing the PV force. Solar power is useful because it is clean, renewable energy.

Companies that produce solar-power devices ensure that the solar panels will consistently and reliably perform for a certain period of time. The dominant makers offer a 25-year standard solar-panel guarantee, which implies that power yield ought not go below 81.04% of appraised control, for at least 25 years.

2.1.4 Monocrystalline Silicon Cells

Monocrystalline silicon cells (single crystalline cells) are used for charging the equipment through solar cells. They are made from pure silicon. They are proven to be most efficient in terms of performance and space as compared to all other solar cells. Their silicon-ingots shape helps to optimizing the performance. The greatest productivity that can be reproduced in the lab with a monocrystalline silicon cell is 25%.

Polycrystalline cells have quite recently passed 20.02% productivity in the lab, while thin-film innovations, for example, copper indium gallium selenide (CIGS) solar cells and cadmium telluride (CdTe) solar cells are just below the 20.02% effectiveness. Indistinct silicon presently has the most reduced effectiveness with around 13.05%.

2.2 Literature Review

Rai et al. [7] proposed a special kind of charger in which there was no need for any kind of wire to connect the power supply to charge it for a short range of distance. This device depletes the use of cables, and users no longer need to sit around the phone and cable just to make it charge. Wireless chargers increase the life span of mobile devices as compared to ordinary chargers. The technology is capable of charging any smartphone, anywhere, at any time. Normally wireless chargers are power-consuming, but in the case of this device, a renewable source of energy is used [8]. In India, there are many towns and villages where electricity is a big concern, so wireless solar chargers are very useful. The device is compatible with all smartphones, as it has very little chance of mobile overcharging and less chance of mobile accidents. It will provides safety from electric shock, making it both user-friendly and environmentally friendly.

Mobile phone charging by the application of a solar-energy harvester has been suggested by Qutaiba [9]. The energy is harvested from the environment without wires. The proposed solar energy harvester is able to charge under indoor light or sunlight efficiently for usages in communication systems and cameras or mobile phones.

Sulaiman et al. [10] presented the design of a wireless charging system to replace the old charging systems. Wireless charging stations for smartphone charging and other electronics applications are being developed at Princeton University as claimed by Neil Savage [11]. In the Netherlands, charging of electric vehicle batteries using solar energy has been presented by Chandra et al. [12]. The optimal location of solar panels has been determined based on Dutch meteorological department data. A new solar-cell junction configuration, known as "Sharon-Schottky," for charging of a unique battery without or with a membrane, has been discussed [6].

MATLAB® software has been used to design a new type of charged controllers [13]. The discharging and charging functions have been confirmed. The solar energy through PV cells depends on the intensity of sunlight. However, the energy required to charge modern electronic devices is very low, which thus enables solar cells to fulfill the demands [14].

A portable solar-panel charger has been designed [15, 16] with an energy density of 571-watt hour/kg. To harvest the electric power, various techniques exist for replacing the mobile batteries [17]. The power can be extracted from outside light, heat, and vibrations.

Current wireless charging systems need to become more practical, efficient, and able to transfer power through air gaps [18]. For remote sensing, solar-power transmission has been proposed by Whitehurst et al. [19]. WPT allows consumers to recharge their mobile phones in the same way as the data are transmitted [20]. Although the demand for renewable energy is continuously on the rise, small energy-harvesting antennas for wearable systems suffer from low efficiency [21]. The environment's energy may be extracted by wireless sensor networks [22] to increase their life span. Fast, low-cost solar chargers working with a vehicle's rooftop PV array have been presented by Nguyen et al. [23]. Wang et al. addressed the multi-round sensor dispatch problem in a hybrid wireless sensor network (WSN) with wireless chargers [24], which contain static and mobile sensors to support various applications. Presented by Rao et al. [3], a PV solar-powered battery charger has been used to control the output power. This chapter presents a solar charger Application-Specific Integrated Circuit (ASIC) for lithium-ion (Li-ion) and nickel-based (NiCd or NiMH) batteries [25]. Renewable energy, including solar energy [26], is cleaner, easy to set up and has a very low cost of maintenance during its operation.

2.3 Proposed Methodology

In the proposed work, the monocrystalline silicon (single crystalline cell) for charging the equipment through solar cells has been used, which is made from a pure type of silicon. It is proven as most efficient as compared to other types of solar cells. Its silicon-ingots shape helps in optimizing its performance. The monocrystalline solar cell is comparatively more expensive than other solar cells, but it lasts longer with low maintenance.

The advantage of the product is that it can charge mobile devices using solar power and WPT. There is no need to carry a charger device and mobile phone all the time. The work is based on natural energy, that is, solar energy by which the device gets charged. Solar power is a renewable energy source that has low maintenance costs with diverse applications. The solar power being used is captured in the monocrystalline solar cells that are most efficient in performance. WPT is combined with these cells to give the best charging experience while attaching the device to the back of the cell phone. This increases mobility while using mobile devices (cell phones). The proposed charging device is based on mutual inductance, magnetic resonance, and inductive coupling on different three-dimensional (3D) images, which are helpful to visualize the device.

2.3.1 Wireless Power Transmission Using Solar Power

The combination of using solar power along with WPT is a good concept. In the proposed work, it is found that when solar energy is combined with electrical energy to be used on mobile phones, the battery of the mobile drains very swiftly. The proposed device helps to charge the mobile device seamlessly without the hassle of wires; when the charger is attached to the back of the smartphone with the help of integrating metal plates, the charger automatically charges the smartphone. Both devices can achieve this mobility with zero electricity consumption at least while charging the mobile devices under sunlight.

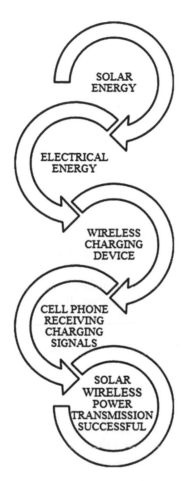

FIGURE 2.3
Framework for wireless power transmission using solar power.

When the devices are paired with the help of metal plates, the sensor chips play their role. When sunlight hits the solar panel, which has monocrystalline silicon cells, the inner portion making solar power will transfer the energy into the device by using sensors. In the proposed device shown in Figure 2.3, energy can also be stored for later use, just like a portable mobile charger. But in this case, that portable charger is getting power from the sun and there are no wires attached to it. This user-friendly device will make mobile charging much more convenient in people's day-to-day lives.

2.3.2 Voltage Rating

The voltage rating of a generator is ordinarily expressed as the working voltage between two of its three terminals, i.e., phase-to-phase voltage. For a twisting associated in delta, this is equivalent to the stage winding voltage. For a twisting associated in wye, it is equivalent to the square root of three times the stage winding voltage. The limit rating of the machine contrasts from its pole management due to two parts, specifically: First is the power factor and the second is productivity. The power factor is that proportion of real power sent to the electrical load partitioned off by the combination voltage. The productivity is that proportion of the electrical power that yields to the mechanical power input. The distinction between these two power esteems is the power suppression comprising of losses in the magnetic iron because of the evolving flux, losses in the opposition of the stator and rotor conductors, and losses from the windage and bearing rubbing. In vast synchronous generators, these losses are under 5.03% of the limit rating. These losses must be expelled from the generator by a cooling framework to maintain the temperature inside the breaking point forced by the protection of the windings.

2.3.3 Controllers

Whenever a panel is utilized that is more than 5-watt evaluated output, it is recommended to use a solar-charge controller. As a matter of fact, a charge controller is a smart thought in a variety of uses, especially for forestalling overcharge, upgrading charge status, and dodging battery discharge in low- or no-light circumstances. Some solar panels are made with blocking diodes pre-installed that hinder battery discharge amid low- or no-light conditions.

For many instances in which a 6-watt or larger solar panel is used, the utilization of a charger controller is strongly recommended. It acts similarly to an on and off switch, facilitating capacity to pass when the battery needs it and cutting off the capacity when the battery is completely charged. A 6-ampere solar-charge controller is sufficient to operate with about every panel, up to and including about 70 watts.

2.3.4 Operating a Device Directly from Solar Panel

A few foldable/compact solar panels that accompany a female cigarette lighter connector can be carried for hiking. Keeping in mind the end goal of interfacing straightforwardly to a panel, the gadget cannot be sensitive to voltage variation, or it may fail. To avoid this issue, it is best to employ a little battery as a space vessel for energy that will give a consistent source of steady and reliable power.

2.3.5 Wireless Charging Technology

Wireless charging works by exchanging the energy from the charger to a recipient attached to the back of a cell phone through electromagnetic acceptance. The charger utilizes an acceptance loop to make a substituting electromagnetic field, which the receiver coil in the cell phone converts again into power to be bolstered into the battery. The charger and cell phone regularly must be close to each other and effectively adjusted to the highest point of each other. Charging at a distance has huge market potential for wireless charging with charging pads.

2.4 Working

The design of proposed solar WPT, including side view, upper side view, bottom view and top view is shown in Figure 2.4. With the help of WPT inside, the hexagonal base will be the magnetic coils. The solar panel here is a wedge in nature in which monocrystalline silicon cells are used. To have two sides for more solar-power utilization, connection of

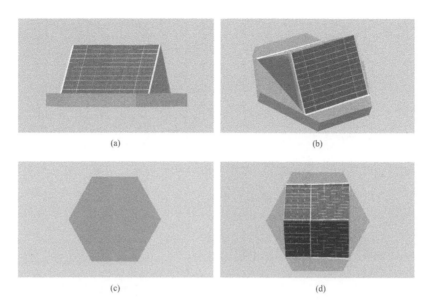

FIGURE 2.4
Solar wireless power transmission: (a) side view, (b) upper side view, (c) bottom view, (d) top view of proposed device.

solar panels to the wireless pad is done inside using sensor chips as it has been merged with the panel and wireless charging pad. After that the wireless charger will be ready.

The base shape is converted into a torus with a rectangular panel and the design of that with its top view, side view and bottom view in Figure 2.5. The panel gets light from the sun and charges the pad simultaneously with the help of sensor chips that

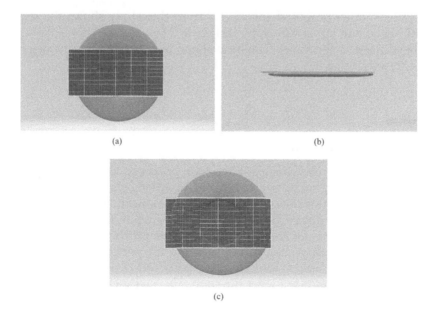

FIGURE 2.5
The base shape is converted into a torus and the panel is in a rectangular shape: (a) top view, (b) side view, (c) bottom view.

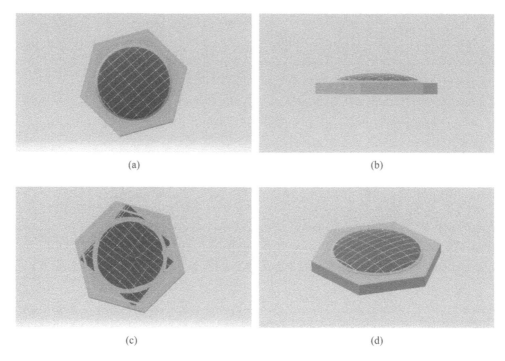

(a)   (b)

(c)   (d)

FIGURE 2.6
The base shape is hexagonal, and the panel is spherical shaped: (a) top view, (b) side view, (c) bottom view, (d) upper view of device.

charge the device when kept on the wireless charging pad integrated with a metal plate on both devices.

The design of another proposed solar WPT from this study shows the top view, side view, bottom view, and upper side view in Figure 2.6(a–d). Two monocrystalline silicon cells have been used in the panel. The base shape is converted into hexagonal and the panel is in a spherical shape.

2.5 Results and Discussions

In this work, IoT-based wireless devices have been proposed to work in conjunction with solar panels, wireless charging pads, and sensor chips along with configuring metal plates to simultaneously charge the devices. This provides connection of each object by means of a WSN. The proposed device conserves electricity while increasing users' mobility. Monocrystalline silicon cells for charging the equipment through solar panels have been used. Their silicon-ingots shape helps in optimizing the performance. The monocrystalline solar cell is expensive as compared to other solar cells, but they require less maintenance and have about 25 years or longer life span. This space-efficient device allows users the flexibility to place the devices or batteries in one or multiple dimensions. Solar power is a renewable energy source that has low maintenance costs with diverse applications. WPT is combined with these cells to give the best charging experience in short range where the mobile device can be freely used. It increases mobility while using the mobile device.

2.6 Conclusion

The proposed work yields one of the most renewable energy sources—solar energy. Existing products are not designed to use solar energy as the source of power energy and do not have any wireless devices especially up to the proposed device range. The prototype is cost-effective and easily accessible by everyone. The complete setup is feasible, reliable, and cost-effective with almost free power consumption. An attempt has also been made to minimize the use of electrical energy and minimize pollution to the environment. The proposed model may provide a good solution, particularly for people in rural areas where regular availability of electricity is always a problem.

References

1. Na, W., Park, J., Lee, C., Park, K., Kim, J., & Cho, S. (2018). Energy-efficient mobile charging for wireless power transfer. *Internet of Things Journal, 5*(1), 79–92.
2. Prawiro, S. Y., & Murti, M. A. (2018). Wireless power transfer solution for smart charger with RF energy harvesting in public area. In IEEE 4th World Forum on Internet of Things (WF-IoT), Singapore, 103–106, doi: 10.1109/WF-IoT.2018.8355160
3. Rao, X., Yang, P., Yan, Y., Liu, G., Zhang, M., & Xu, W. (2017). Optimal deployment for roadside wireless charger with bounded detouring cost. In ICC2017: WS02-IEEE Workshop on Emerging Energy Harvesting Solutions for 5G Networks (5G-NRG), 493–497.
4. Chieochan, O., Saokaew, A., & Boonchieng, E. (2017). Internet of things (IoT) for smart solar energy: A case study of the smart farm at Maejo University. In 2017 International Conference on Control, Automation and Information Sciences (ICCAIS), 262–267, doi:10.1109/ICCW.2017.7962706.
5. Bardwell, M., Wong, J., Zhang, S., & Musilek, P. (2018). Design considerations for IoT-based PV charge controllers. In IEEE World Congress on Services, 59–60, 10.1109/services.2018.00043
6. Maheshwar, S., Veluchamy, P., Natarajan, C. and Kumar, D. (1991). Solar rechargeable battery—principle and materials. *Electrochimica Acta, 36*(7), 1107–1126.
7. Rai, H. M., Sisodiya, P., & Agrawal, I. (2016). Design of solar powered wireless charger for smartphones. *IRJET Journal, 3*(5), 1278–1281.
8. Srinivas, P., & Vijaya Lakshmi, K. (2017). Solar energy harvester for wireless sensor networks. *International Journal of Innovative Research in Electrical, Electronics, Instrumentation and Control Engineering, 5*(6), 240–247.
9. Qutaiba, A. (2011). Design and implementation of a mobile phone charging system based on solar energy harvesting. *Iraq Journal Electrical and Electronic Engineering, 7*(1), 264–267.
10. Sulaiman, N., Shawal, M., Abdul, J. A., Muhammad, H., Mohd, S. N., Ahmad, A., … Yusuf, A. (2016). Evaluation of mobile phone wireless charging system using solar and inductive coupling. In *Advances in Information and Communication Technology, 538,* 238–247.
11. Savage, N. (2012). Wireless solar charging made easier. IEEE Spectrum. https://spectrum.ieee.org/green-tech/solar/wireless-solar-charging-made-easier
12. Chandra Mouli, G. R., Bauer, P., & Zeman, M. (2016). System design for a solar powered electric vehicle charging station for workplaces. *Journal of Applied Energy, 168,* 434–443.
13. Reddy, M. L., Kumar, P. J. R. P., Chandra, S. A. M., Babu, T. S., & Rajasekar, N. (2017). Comparative study on charge controller techniques for solar PV system. *Energy Procedia, 117,* 1070–1077.
14. Schubert, M. B., & Werner, J. H. (2006). Flexible solar cells for clothing. *Materials Today, 9*(6), 42–50.

15. Yau, H-T., Lin, C-J., & Liang, Q-C. (2013). PSO based PI controller design for a solar charger system. *The Scientific World Journal-Hindawi*, *2013*(3):815280. https://www.hindawi.com/journals/tswj/2013/815280/abs/

16. Gadelovits, S., Sitbon, M., Suntio, T., & Kuperman, A. (2015). Single-source multibattery solar charger: Case study and implementation issues. *Progress in Photovoltaics*, *23*(12), 1916–1928.

17. Paradiso, J. A., & Starner, T. (2007). Energy scavenging for mobile and wireless electronics, *IEEE Pervasive Computing*, *4*(1), 18–27.

18. Beh, T. C., Kato, M., Imura, T., Sehoon, O., & Hori, Y. (2013). Automated impedance matching system for robust wireless power transfer via magnetic resonance coupling. *IEEE Transactions on Industrial Electronics*, *60*(9), 3689–3698.

19. Whitehurst, L. N., Lee, M. C., & Pradipta, R. (2013). Solar-powered microwave transmission for remote sensing and communications. *IEEE Transactions on Plasma Science*, *41*(3), 606–612.

20. Tomar, A., & Gupta, S. (2012). Wireless power transmission: Applications and components. *International Journal of Engineering Research & Technology*, *1*(5), 1–5.

21. Kansal, A., & Srivastava, M. (2003). An environmental energy harvesting framework for sensor networks. In Proceedings ACM International Symposium on Low Power Electronics and Design, 481–486.

22. Khan, S., Ahmad, A., Ahmad, F., Shemami, M. S., Alam, M. S., & Khateeb, S. (2017). A comprehensive review on solar powered electric vehicle charging system. *Smart Science*, *6*(1), 54–79.

23. Nguyen, T-T., Kim, H-W., Lee, G-H., & Choi, W. (2013). Design and implementation of the low cost and fast solar charger with the rooftop PV array of the vehicle. *Solar Energy*, *96*, 83–95.

24. Wang, Y-C., & Huang, J-W. (2018). Efficient dispatch of mobile sensors in WSN with wireless chargers. *Pervasive and Mobile Computing*, *51*(4), 1–13.

25. Pastre, M., Krummenacher, F., Kazanc, O., Pour, N. K., Pace, C., Rigert, S., & Kayal, M. (2011). A solar battery charger with maximum power point tracking. In 18th IEEE International Conference on Electronics, Circuits, and Systems, Beirut, 394–397, 10.1109/ICECS.2011.6122296.

26. Thakur, T. (2016). Solar power charge controller. *Global Journal of Researches in Engineering*, *16*(8), 1–13.

3

Internet of Things and Its Applications

Pradeep K. Garg[1] and Lavanya Sharma[2]

[1]*Civil Engineering Department, Indian Institute of Technology, Roorkee, India*

[2]*Amity Institute of Technology, Amity University, Noida, Uttar Pradesh, India*

CONTENTS

3.1 Introduction

The IoT has essentially increased the capabilities of remote distance control using a diversity of interconnected things or devices. The network of several connected devices is called IoT [1]. For example, wireless technology allows for machine-to-machine (M2M) enterprises to monitor and operate their equipment. For this purpose, proprietary standards were initially used, then the Internet protocol (IP) and Internet standards. The use of IP to connect devices, and research and development into smart-object networking gave birth to IoT. Figure 3.1 presents the idea behind the development of IoT. Here, the A's refer to the global utilization of the technology (anytime, anywhere, any device, any device, any network, etc.) and the C's indicate the properties of IoT, such as collections, convergence, connectivity, and computing. IoT today has already developed much beyond the A's and C's [2].

The IoT reduces the information gap between the world and the Internet. The IoT encompasses several uses from connected vehicles to connected houses, tracking an individual's behavior, and monitoring activity. The IoT can be considered a universal global neural network in the space that will cover every aspect of our lives. From a technology point of view,

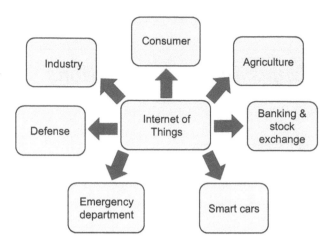

FIGURE 3.1
Applications of Internet of Things.

the IoT can be defined as smart objects, machines interacting and communicating with other machines using sensors, and smart environments and infrastructures, resulting in enormous amounts of data generated, which are processed and converted into useful activities that can command and control the things and make living much more comfortable and convenient.

The IoT has wide applications in the fields of infrastructure, traffic and transportation, lifestyle, smart building, smart homes, smart agriculture, smart environment, smart security, factory, healthcare, and many more. For example, utility companies can use an IoT system for energy products, such as the billing process. With developments in networking and computing, the Internet, improved sensors, software, and wireless networks have encouraged the use and applications of IoT systems in many tasks across the world.

3.2 IoT Technology

The IoT comprises various components. The physical objects or devices, also known as *things*, can easily sense the environment by means of a network of sensors. Human beings are required for interaction, for example in building automation, as the environment can be controlled through mobile applications. The platforms are used to connect IoT components with the physical objects and IoT. These objects are connected via networks using different means of communication technologies [1]. Huge volumes of data, thus gathered, are processed and converted into valuable information that may be required by various applications running on the IoT components.

Due to a network of networks, IoT can perform various tasks efficiently and accurately. Geographic information system (GIS) programs within the IoT are being implemented more broadly; be it smart homes, cars, or transportation. The data generated through GIS analysis can yield automatic reports, and greatly assist businesses in making relevant decisions. The software can convert these data into actionable insights and improve the operational efficiency.

Sensors are used in most places, and now sensor technology is guiding and helping human beings [3]. This is made possible through a technology called *sensor fusion*. Sensor fusion fuses individual data collected from several sensors in order to obtain highly accurate data then obtained from the data from each individual sensor. It helps in context awareness, having good potential for IoT. The sensors can be used to provide services to people. Sensor fusion provides aggregate information, which is much greater, more accurate, and more reliable than information obtained from an individual sensor.

In the IoT field, sensor fusion plays a similar role. By integrating information from several sensors, sensor fusion can provide much higher levels of recognition. Individual sensors that have limitations can be improved by complementary sensing nodes. For different IoT applications, the sensing nodes may vary widely. Sensing nodes mainly include a camera system for monitoring of images, and electricity flow meters for smart energy. For example, radio frequency identification (RFID) readers sense the existence of a particular object or person, doors and locks with circuits indicate intrusion in the building, or a digital thermometer can measure the temperature [4]. These nodes have a unique ID, and can be controlled separately via a remote command. RFID will have a large number of applications, such as remote command and control of the IoT.

IoT-based tracking and monitoring of any resource is economical for companies and people. For this reason, the IoT is going to have a significant major impact on our lives. But, it is important to understand the advantages and disadvantages of IoT, before it is adopted for an application.

3.3 Human Beings: The Ultimate Sensing Example

The brain is the ultimate decision maker. The brain uses inputs from several sensors to validate an event and to make a decision. We need to understand how sensor fusion works in a human body. A human being can feel the external environment in several ways. For example, hearing, vision, sensations of smell and taste, and sensation of touch, all provide sensory information about the external environment, which reaches the brain through the peripheral nervous system. The brain then makes a decision to respond to a given situation. For example, if a driver wants to overtake a slow-moving car, his/her brain guides his/her muscles to drive faster to overtake and then slow down a little after he/she passes the slow car. In this example, the information traveling to the brain to guide decisions and react is much greater than the sum of the disparate sensory inputs.

The peripheral nervous system cannot make complex decisions about the information it collects and transports. The brain makes these decisions. Human beings would not be able to perform many functions without the ability of the peripheral nervous system to bring in sensory information to the brain. The internal organs of human beings also receive information, which is sometimes noticeable, such as a headache. So, it is evident that human beings have a sensory system that helps in much functionality.

3.4 Applications of IoT

There is large number of applications of IoT, worldwide, as shown in Figure 3.2. Table 3.1 presents a summary of IoT applications globally [5]. It is practically impossible to list all potential IoT applications looking at the developmental needs of potential users. In the following sections, some of the important and common applications are summarized.

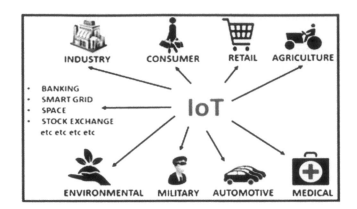

FIGURE 3.2
Applications of Internet of Things.

TABLE 3.1

Global Share of IoT Projects in Various Segments

S.No	IoT Segment	Global share of IoT projects	US	Europe	APAC Countries
1	Smart city	20	31	47	15
2	Smart energy	13	49	24	25
3	Smart agriculture	6	48	31	17
4	Smart retails	4	52	30	13
5	Smart supply chain	4	57	35	4
6	Connected buildings	5	48	33	12
7	Connected car	13	43	33	17
8	Connected health	5	61	30	6
9	Connected industry	22	43	30	20

3.4.1 Agriculture

Agriculture is the most important core industry in the world. According to the Food and Agriculture Organization of the United Nations [6], food production has to be increased by 60% by the year 2050 to meet the demand of the world's growing population, which is expected to reach 9 billion. Increased crop productivity is a global challenge that may include smart farming, which involves the use of information communication technologies (ICT), IoT, and big data analytics to take up these challenges by monitoring the soil, moisture, and crops, as well as related environmental parameters. Precision agriculture and related technologies offer various management options to optimize the variable rates of fertilization, but the increasing demand for fertilizer may have negative impacts on water, air, soil, and climate that can also adversely affect crop quality [7].

The range of IoT applications in agriculture is large. One of the popular applications of IoT in agriculture is precision farming with the adoption of smart technology that includes sensors, robots, systems of control, and autonomous vehicles. In addition, the potential of IoT in farming includes offering solution to farmers to apply environmentally friendly pesticides. Previously, agriculture was fully dependent on human resources and hard machinery, but it has now started applying technological solutions and modernizing its various operations with IoT applications that are intended to make a significant impact on agriculture. In addition to the introduction of smart technology into agriculture management, IoT also enables proper tracking of natural factors, such as climate change, soil composition, and weather forecasts.

The European Space Agency under the framework of the Integrated Applications Programme (IAP), carried out a project on space-based services integrating the IoT [8]. The Spacenus system aims to address agricultural plant nutrient deficiency (PND). The efficiency of fertilizer use highly depends on PNDs and field nutrient budgeting. In Spacenus, satellite-based fertilization maps can provide information about the relative spatial variability of the crop nutrient status. Using these maps, farmers can make decisions about specific fertilization for the best economic return. A camera/sensor, which is connected to an artificial intelligence (AI) cloud, can capture, analyze, and provide actionable information in real-time. The system is powered by a deep convolutional neural network that can recognize crop-specific plant diseases with a high degree of accuracy (96%).

Through the machine data link services of Spacenus, farmers can easily translate fertilization maps into machine-readable application maps and transfer the maps into their tractors. Photographs of the field crops can be taken with a smartphone. The Spacenus system can detect and quantify the PND, and that information enables the farmers to diagnose the crop nutrient deficiency in a quantitative manner. By combining this in-situ PND information with the satellite-derived fertilization map, farmers can easily derive a variable-rate fertilization map.

Precise fertilization to the crops requires knowledge on PND. The main objective of IoT-based application service is monitoring PND and subsequently providing real-time information on fertilization necessities.

A multidisciplinary Australian team [9] developed a system called SmartFarmNet. The SmartFarmNet technology is considered to be the largest system in the world to provide crop performance analysis and recommendations. The unique system allows users to utilize the IoT sensor of their choice, and features effortless integration with any IoT device, which reduces sensor installation and maintenance costs, and supports scalable data analytics. SmartFarmNet also provides tools for faster analysis of big data, which is generated from large numbers of IoT sensors.

3.4.2 Supply Chain in Agriculture

To enhance agricultural productivity, the satellite imagery and intelligence IoT for agriculture (SIITAg) tool can identify PND and generates a variable-rate fertilization application map to get optimal agricultural output [10]. To manage fertilization demands, farmers require tailor-made advice suited to their fields. SIITAg generates a variable-rate fertilization map for each zone based on crop type and crop condition, which is essentially needed by farmers to attain the optimum nutrient concentration. The system is economical, scalable, and easily available to farmers worldwide. Initially, the service will be used in Germany, and later it will be available for the EU countries, and the US.

Several companies require state-of-the-art technology that is integrated with easy-to-use interfaces so that their customers are provided with valuable additional services as compared to other companies. They also like to monitor how the demand develops among their customers so that their supply-chain and marketing strategies are effectively designed.

SIITAg uses satellite images from Sentinel-2 satellite as well as Galileo global navigation satellite system (GNSS) observations for computational purposes. The SIITAg front end operates as a smartphone app, while the back-end system employs the capabilities of AI. A machine-learning algorithm combines the satellite data with data collected through the app to create the recommendations [10]. Farmers can see their fertilizer needs from fertilization recommendation maps using the SIITAg service. The farmers take images of some of the crop leaves in the field with their smartphone's app, which are analyzed by image-processing algorithms to quantify the exact plant nutrient contents (N, P, K, S, and Mg). The output map can be uploaded directly to the farmer's fertilizer spreader, which allows farmers to assess their area's performance and to enhance their decision-making processes.

End-to-end farm management systems are areas of the interest of IoT developers in farming markets. In terms of logistics, IoT-based agriculture enables using global positioning system (GPS), RFID, and other location-based sensors to control the transportation and storing of food plants. By using IoT, the entire supply chain can increase its effectiveness, allowing for improvements in terms of transparency and customer awareness. Sensors and devices can be installed in the system.

Agriculture-based applications can partner with IoT-based agriculture to provide farmers with data for analytics, reports, and accounting [5]. An application called FarmLogs consists of software related to the agriculture market for facilitating grain marketing decisions. It provides a toolkit necessary for creating a grain marketing plan (with the value of unsold crops, contact lists, and goal-setting) and insights on profitability increasing. Among the products, farmers can order marketing, reports, automatic movement recording, crop health imagery, and rainfall tracking. Cropio software provides a system for field management and vegetation control, in addition to the abilities to provide field history, instant alerts, vegetation maps, and soil, moisture, and harvest forecasts. It can maintain a record of the state of numerous fields, provides real-time data on the necessary updates, and assists in forecasting.

3.4.3 Greenhouse Farming

IoT applications in agriculture provide various effective solutions for greenhouse farming, as climate control can be achieved through positioning several sensors that send alerts about water or air problems [5]. The products, such as Farmapp and Growlink allow achieving these aims. Farmapp offers farmers an integrated pest-management software with monitoring, sensors, and fumigation functions. It includes a scouting app for fast recording and implementation of required measures with satellite maps, comparative maps, charts, and reports. Real-time data on weather and soil conditions can be assessed through satellite images and algorithmic calculations. The functionality of Farmapp captures better irrigation, thus IoT in agriculture enables tracking of the amount of water used in plants for its optimization. Growlink allows a real-time monitoring system in greenhouses with the aim to increase quality and yield performance. This IoT agriculture solution particularly concentrates on automation of working with operational data, including planning, controlling, tracking, and monitoring activities. Hence, farmers can achieve the best possible performance in the long run.

Pallavi et al. [11] present remote-sensing and IoT-based agriculture parameters and control system to the greenhouse agriculture. The aim is to control CO_2, soil moisture, temperature, and lighting conditions for the greenhouse crops, and to enable organic farming with increased yields. In the IoT-based approach, information about greenhouse atmosphere parameters, such as CO_2, temperature, and light, is detected, which is then directed to the cloud, and thereafter the desired action is taken by the farmers. It is feasible to control and monitor the precise application of water flow control, regulated temperature range, light radiation, and so on, for the better growth of plants. The system is expected to help farmers to increase yields with minimum visits to the field.

3.4.4 Livestock Management

As livestock control, IoT in agriculture assists in tracking the state of the herd. The applications, such as SCR by Allflex and Cowlar, determine the health of animals, find their locations, and track the state of pregnancies, especially while dealing with cattle and chickens [12]. SCR by Allflex offers cow, milking, and herd intelligence, along with several other professional solutions. Its functionality includes tracking all the insights about each herd participant (heat, health, and nutrition), optimizing the milking process (simplify and streamline), and collecting data into an integrated and actionable plan for herd development. Likewise, Cowlar makes smart collars for dairy cows that address the similar needs for optimizing milking, maximizing performance, and reducing labor costs along with boosting reproduction.

In addition, Symphony Link is an application that avoids mesh networking to address the challenge of the need for monitoring animal health in the long run and accomplishes the task of a complete integration effectively [13]. As a revolutionary invention in the world of IoT in agriculture, it links wide-area IoT networks with modules, gateways, and conductors.

3.4.5 Smart Cities

A smart city operates and provides its services (e.g., governance, safety, transport, traffic, energy, waste management) through the use of information and communications technologies [14]. It can improve the lives of citizens by using advanced technologies to create smart or intelligent services for its citizens. These services might include smart transportation systems, environmental monitoring, air pollution monitoring, or energy optimization [15]. For example, the European smart city project has identified six important characteristics for a smart city, as shown in Figure 3.3. The IoT-based technologies will play an important role by providing additional valuable input to such indicators [16].

By 2020, the development of mega city corridors and networked, integrated, and branded cities is expected. With more than 60% of the world population expected to live in urban cities by 2025, growth in urbanization is expected to influence human lives

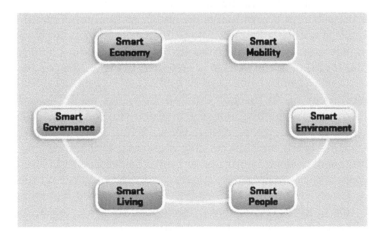

FIGURE 3.3
Six smart characteristics to enhance quality of life [16].

and mobility. Rapid expansion of city borders, along with an increase in population and infrastructure development, would compel city boundaries to further expand outward and merge peripheral small towns to form mega cities, each with a population of more than 10 million. By 2023, there will be 30 mega cities globally, with 55% in developing economies of India, China, Russia, and Latin America [15]. It is expected that there will be about 40 smart cities globally by the year 2025. The smart city will also focus on utilities and environment in these cities and addressing the role of IoT in safety, environmental planning, waste management, water management, lighting, and transport systems [17].

3.4.6 Asset Monitoring

The range of sensors and resolution of remote sensing images available from satellites is growing rapidly to map, monitor, and assess the assets on earth's surface. The satellite application Catapult, based at the Harwell Campus in Oxfordshire, is using satellite remote sensing and IoT technology that can provide a reliable and sustainable solution for improving and lowering asset management costs [18].

Real estate is a key indicator of growth that can be monitored using IoT and associated things. Orbital Insight can provide details about how quickly houses are being built to determine growth in different regions [19]. The data can also help insurance companies to better protect their customers, and also carry out damage assessment from natural disasters. It can help to guide better decisions about resource allocation. Large companies can effectively make use of Orbital Insight because they are making decisions over large geographies. It also applies to more localized individuals, such as farmers who are looking for crop insurance.

In remote and potentially hazardous areas, monitoring civil engineering construction projects particularly poses a challenge for satellite and M2M/IoT [20]. In many industrialized countries, the civil engineering industry is in a transition phase using modern tools from new construction to maintain the stock of valuable assets (e.g., bridges, tall structures, and high-rise buildings). Structural health monitoring systems integrating the satellite technologies and IoT can be a potential means for maintenance of these assets. The system can give a complete picture of the structure to locate threats caused by environmental degradation, landslides, mining, and construction industrial activity, if any. Due to rapidly aging infrastructure worldwide, the ability to monitor structures in hazardous and inaccessible areas will be a key advantage. Newer IoT technologies are being developed to continuously monitor infrastructure such as roads, bridges, and highways. There is a great potential for IoT, and the market is projected to grow 25% per year to $3.4 billion by 2022 [20].

The state government of Arunachal Pradesh designed an IoT-based solution with SkyMap Global, which uses satellite imagery, Internet, and the power of a smartphone to help monitor infrastructure developments in the state [21]. The hilly terrain of Arunachal Pradesh with winding roads and inaccessible areas makes it impossible to monitor the physical development of projects by ground survey. The developed IoT-based system aims to monitor the progress of construction at sites, and sends the actual report to the state government in a timely manner. A GPS-enabled android-based smartphone application is used for collection of information from project work sites. The application is useful for reporting the field conditions, actual images, latitude and longitude coordinates, archives, and relaying field information. Photographs of work sites during progress are also added to the project information.

3.4.7 Traffic Optimization

IoT-based systems include managing the road traffic system by way of intelligent transport systems (ITSs), automated drive vehicles, and automated adaptive traffic control. The objective of an ITS is to optimize and control the traffic for environmental sustainability of the cities. An IoT-enabled ITS would help citizens reach their destinations faster and safer. It also helps to make better decisions in selecting the optimal route with optimal driving speeds [1].

Traffic regulation could be another important IoT application. Various sensors prefixed on the road sides can monitor traffic's volume and speed. These sensors can transfer the data to a system using algorithms to optimize the traffic flow. With the integration of Wi-Fi into fleet vehicles, drivers and managers are now aware of real-time sharing of vehicle information, such as GPS tracking and routing information. Positioning and visualization technologies are already in use for autonomous vehicles, and they may be used for other transportation applications.

The application of IoT in geospatial contexts (remote sensing images, GPS) is also useful in creating time-related data/maps of potholes in roads, and noise around a metropolitan city [22], with classification showing seasonal changes. With geospatial tools, it is also possible to monitor the daily travel routines and activity patterns of commuters. The use of IoT thus can help traffic to improve efficiency through Wi-Fi and GPS tracking.

3.4.8 Healthcare

Healthcare can benefit in two ways; namely through preventive intervention as well as through remote monitoring. Earlier, hardly any medical sensing technology, except thermometers, was available outside of hospitals. Sensors allow not only efficient, but also constant, remote monitoring of health data, including times when people show healthy symptoms. Without data tracking, an illness often remained undetected until it broke out. Now with progress in preventive healthcare, data helps us to control diseases at an early stage. The longer the time duration of recording individual health data, the more these data empower the experts to compare and identify irregularities and then to provide preventive, personalized treatment [23].

The HealthMap project [24] analyzes such incidents to improve patients' disease risk at the local level and to suggest proper prevention measures. Sensor systems are important tools for maintaining a clean, safe, and healthy environment that can detect the presence of and quantify the gases present in the environment. Once the source is located, monitoring by sensors can support mitigation and compliance. For example, heat maps are commonly used for visualizing urban air quality [25], and monitoring of dengue fever [26].

The IR sensors are gaining popularity in IoT M2M applications, including medical diagnostics, imaging and industrial process controls, fire detection and gas leak detection, pollution source and monitoring, and so on. IoT applications for devices using IR sensors include home-entertainment products, medical/healthcare equipment, home appliances with temperature control, movement detection, and body-temperature measurements.

Telemedicine is also another IoT application that has gained popularity in recent years. In another application of Iot, heart patients are asked to wear special monitors that are designed to gather data or information related to the heart; signals from the monitor are forwarded to the patient's implanted device so the device can deliver a timely shock to correct an errant heart rhythm.

A large number of health-based services can be improved by introducing geospatial-based systems such as disease prevention, initial warning, diagnostics, and treatment [27]. Services, infrastructure, and facilities provided by satellite communications or satellite navigation can also be utilized for new applications to provide health-related solutions to society as a whole.

3.5 Conclusion

The IoT has already impacted numerous fields positively. Today, IoT is used in several areas, including asset monitoring, security, fleet tracking and management, energy management, and many more. Sensors are used in a wide range of IoT-based applications, such as smart mobile devices, automotive systems, healthcare, industrial control system, disaster management, and weather monitoring.

By combining all of these technologies, the output from the various sensors is much more useful than its separate fragments. Sensor fusion or networks of sensors provide capabilities to make our lives simpler and easier and also enable a variety of health-related services that require these capabilities. One of the challenges being faced by the industry today is lack of standardization in sensors across various operating systems. Today, most operating system drivers ask for the most basic sensor data, which limits the use of the full capabilities of the sensors for providing optimum data for the application at hand.

References

1. *White Paper, Internet of things (IoT): Technology, economic view and technical standardization, version 1.0.* (2018, July). Agence pour la Normalisation etl'Economie de la Connaissance (ANEC).
2. Gupta, R. (2016). ABC of Internet of things: advancements, benefits, challenges, enablers and facilities of IoT. In IEEE Symposium on Colossal Data Analysis and Networking (CDAN).
3. Gubbi, J., Buyya, R., Marusic, S., & Palaniswami, M. (2013). Internet of Things (IoT): A vision, architectural elements, and future directions. *Future Generation Computer Systems, 29*(7), 1645–1660.
4. Said, O., & Tolba, A. (2016). Performance evaluation of a dual coverage system for Internet of things environments. *Mobile Information Systems, 2016*(1), 1–20. http://dx.doi.org/10.1155/2016/3464392
5. Stokes, P. (2018). IoT applications in agriculture: The potential of smart farming on the current stage. https://medium.com/datadriveninvestor/iot-applications-in-agriculture-the-potential-of-smart-farming-on-the-current-stage-275066f946d8
6. FAO. (2016). http://www.fao.org/home/en/
7. Lakshmi, K., & Gayathri, S. (2017). Implementation of IoT with image processing in plant growth monitoring system. *Journal of Scientific and Innovative Research, 6*(2), 80–83.
8. Satellite Imagery and Intelligent IoT for Agriculture (SIITAg) –ESA-IAP Project Kick-off. (2017, October 6). https://spacenus.com/news-1/2018/4/21/satellite-imagery-and-intelligent-iot-for-agriculture-siitag-esa-iap-project-kick-off
9. Jayaraman, P. P., Yavari, A., Georgakopoulos, D., Morshed, A., & Zaslavsky, A. (2016). Internet of things platform for smart farming: Experiences and lessons learned. *Sensors, 16*(11), 1884, doi:10.3390/s16111884

10. SIITAg—Satellite-enabled Intelligent Internet of Things for Agriculture. (2018, January). https://business.esa.int/projects/siitag

11. Pallavi S., & Jayashree, D. (2017). Remote sensing of greenhouse parameters based on IoT. *International Journal of Advanced Computational Engineering and Networking, 5*(10).

12. www.allflex.global

13. https://www.link-labs.com/symphony

14. Borgia, E. (2014). The Internet of things vision: Key features, applications and open issues. *Computer Communications, 54,* 1–3.

15. Frost & Sullivan Inc. Mega trends: Smart is the new green. http://www.frost.com/prod/servlet/our-services-page.pag?mode=open&sid=230169625

16. http://www.smart-cities.eu/

17. Outsmart, FP7 EU project, part of the Future Internet Private Public Partnership. OUTSMART-Provisioning of urban/regional smart services and business models enabled by the Future Internet. http://www.fi-ppp-outsmart.eu/enuk/Pages/default.aspx

18. https://sa.catapult.org.uk/

19. https://orbitalinsight.com/

20. Satellite technologies for IoT applications. (2017, March). Report produced by IoT UK.

21. Infrastructure Monitoring—A Case Study. (2018, June 19). https://skymapglobal.com/infrastructure-monitoring-with-satellite-imagery-a-case-study/

22. Maisonneuve, N., et al. (2009, May 17-21). Citizen noise pollution monitoring. In Proceedings of the 10th Annual International Conference on Digital Government Research, Puebla, Mexico, pp. 96–103.

23. Ziesche, S., & Motallebi, S. (2013). Personalized remotely monitored healthcare in low-income countries through ambient intelligence. *Communications in Computer and Information Science, 413,* 196–204.

24. Brownstein, J., Freifeld, C., Reis, B., & Mandl, K. (2008). Surveillance sans frontieres: Internet-based emerging infectious disease intelligence and the HealthMap project. *PLoS Med, 5*(7), e151.

25. Kotsev, A., Schade, S., Craglia, M., Gerboles, M., Spinelle, L., & Signorini, M. (2016). Next generation air quality platform: Openness and interoperability for the Internet of things. *Sensors, 16*(3), E403.

26. Delmelle, E., Zhu, H., Tang, W., & Casas, I. (2014). A web-based geospatial toolkit for the monitoring of dengue fever. *Applied Geography, 52,* 144–152.

27. Elhadj, H. B., Elias, J., Chaari, L., & Kamoun, L. (2016). A priority based cross layer routing protocol for healthcare applications. *Ad Hoc Networks, 42,* 1–18.

Part 2

Tools and Technologies of IoT with Visual Surveillance

4

Visual Surveillance of Human Activities: Background Subtraction Challenges and Methods

Thierry Bouwmans[1] and Belmar García-García[2]

[1]*Laboratoire MIA, University of La Rochelle, La Rochelle, France*

[2]*Instituto Politécnico Nacional, Mexico City, Mexico*

CONTENTS

4.1 Introduction

The increase in video camera usage has led to background subtraction being used more and more in different visual surveillance applications of human activities, which involve different uncontrolled environments and types of targets to be detected. In video surveillance systems, the main objectives are the identification and tracking of objects. Traffic scenes are the most common environments for the detection of incidents such as vehicles stopped on roads [1–4] or to monitor vehicular traffic on highways, classifying them as empty, fluid, heavy, or with traffic jams. This makes it necessary to detect, track, and count vehicles [5–7]. The background subtraction approach can also be applied in congestion in urban traffic areas, for the detection of illegal parking [8–12], and for the detection of free parking places [13–15]. Safety at train stations and airports is another area of interest because baggage theft is a frequent problem.

Within maritime scenes, surveillance can include counting the number of ships, as well as detecting and tracking vessels in fluvial channels. Other settings are stores, that is, the detection and monitoring of customers [16–19]. Background subtraction also plays an important role in rapid decision-making in sports activities. For example, Hawk-Eye software has become a key part of soccer and tennis contests. Background subtraction can be used for precise analysis of athletic performance, because it has no physical effect on the athlete as in Tamas et al. [20] for rowing motions, and for surveillance as in Bastos [21] for activities of surfers. In this chapter, we thus attempt to survey the main visual surveillance applications of human activities that use background subtraction in their process by classifying them in terms of aims, environments, and objects of interest. We reviewed only the publications that specifically address the problem of background subtraction in the concerned applications with experiments on corresponding videos. Thus, this chapter refers only to background subtraction methods that are previously and currently used in these applications. To have an overview of all the background subtraction methods in research, the reader can refer to numerous surveys [22–29] in the field.

4.2 Background Subtraction: A Short Overview

In this section, we discuss the aim of background subtraction for discerning static and moving foreground objects from a video stream. This task is the fundamental step in many visual surveillance applications for which background subtraction offers a suitable solution that provides a good compromise in terms of quality of detection and computation time. The different steps of the background subtraction methods are as follows:

1. **Background initialization** refers to the generation or extraction of the first background image.
2. **Background modeling** refers to the mathematical model to describe the background. It is also called background representation.
3. **Background updating** consists of updating the background model over time in the video sequence.
4. **Pixel classification in the background/foreground** consists of foreground detection, that is, pixel classification in the class "background" or the class "moving objects."

These different steps are reviewed by Bouwmans [28]. Practically, these steps require methods that have different aims and constraints. Thus, they need algorithms with different features. Background initialization requires offline algorithms that are batched by taking all the data at one time. On the other hand, background maintenance needs online algorithms that are incremental, that is, they take the incoming data one by one. Background initialization, modeling, and maintenance require reconstructive algorithms while foreground detection needs discriminative algorithms. Thus, the background subtraction process includes the following stages: (1) the background initialization module provides the first background image from N training frames, (2) foreground detection that consists of classifying pixels as foreground or background is achieved by comparing the background image and the current image, (3) the background maintenance module updates the background image by using the previous background, the current image, and the foreground detection mask. Steps 2 and 3 are executed repeatedly as time progresses. It is important to see that two images are compared: The methods compare a subentity of the entity background image with its corresponding subentity in the current image. This subentity can be the size of a pixel, a region, or a cluster. Furthermore, this subentity is characterized by a *feature,* which can be a color feature, edge feature, texture feature, stereo feature, or motion feature [30, 31]. In developing a background subtraction method, researchers must design each step and choose the features in relation to the challenges they want to handle in the concerned applications. In the following section, we carefully review the challenges associated with intelligent visual surveillance of human activities.

4.3 Intelligent Visual Surveillance of Human Activities

Surveillance of human activities is the main goal of the background subtraction approach. This section analyzes only articles after the year of 1997, because older detection techniques were based on the difference of a few frames due to the limitations of computation processing. Table 4.1 groups the main applications of visual surveillance of human

TABLE 4.1

Visual Surveillance of Human Activities

Subcategories—Aims	Objects of Interest—Scenes	Authors—Dates
1) Road Surveillance	**1-Cars**	
1.1) Vehicle detection		
Vehicle detection	Road traffic	Zheng et al. (2006) [34]
Vehicle detection	Urban traffic (Korea)	Hwang et al. (2009) [43]
Vehicle detection	Highway traffic (ATON Project)	Wang and Song (2011) [44]
Vehicle detection	Aerial videos (US)	Reilly et al. (2012) [45]
Vehicle detection	Intersection (CPS) (China)	Ding et al. (2012) [46]
Vehicle detection	Intersection (US)	Hao et al. (2013) [47]
Vehicle detection	Intersection (Spain)	Milla et al. (2013) [48]
Vehicle detection	Aerial videos (VIVID Dataset [41])	Teutsch et al. (2014) [35]
Vehicle detection	Road traffic (CCTV cameras) (Korea)	Lee et al. (2014) [49]
Vehicle detection	CD.net dataset 2012[44]	Hadi et al. (2014) [50]
Vehicle detection	Northern Jutland (Danemark)	Alldieck (2015) [51]
Vehicle detection	CD.net dataset 2012 [44]	Aqel et al. (2015) [52]
Vehicle detection	CD.net dataset 2012 [44]	Aqel et al. (2016) [53]
Vehicle detection	CD.net dataset 2012 [44]	Wang et al. (2015) [54]
Vehicle detection	Urban traffic (China)	Zhang et al. (2016) [55]
Vehicle detection	CCTV cameras (India)	Hargude and Idate (2016) [56]
Vehicle detection	Intersection (US)	Li et al. (2016) [57]
Vehicle detection	Dhaka city (Bangladesh)	Hadiuzzaman et al. (2017) [58]
1.2) Vehicle Detection/ Tracking		
Vehicle detection/Tracking	Urban traffic/Highway traffic (Portugal)	Batista et al. (2006) [6]
Vehicle detection/Tracking	Intersection (China)	Qu et al. (2010) [59]
Vehicle detection/Tracking	Urban traffic (China)	Ling et al. (2014) [36]

(Continued)

TABLE 4.1 *(Continued)*

Visual Surveillance of Human Activities

Subcategories—Aims	Objects of Interest—Scenes	Authors—Dates
Vehicle detection/Tracking	Highway traffic (India)	Sawalakhe and Metkar (2014) [60]
Vehicle detection/Tracking	Highway traffic (India)	Dey and Praveen (2016) [61]
Multi-vehicle detection/ Tracking	CD.net dataset 2012 [44]	Hadi et al. (2017) [5]
Vehicle tracking	NYCDOT video/NGSIM US-101 highway dataset (US)	Li et al. (2016) [62]
1.3) Vehicle Counting		
Vehicle counting/Classification	Donostia-San Sebastian (Spain)	Unzueta et al. (2012) [37]
Vehicle detection/Counting	Road (Portugal)	Toropov et al. (2015) [63]
Vehicle counting	Lankershim Boulevard dataset (US)	Quesada and Rodriguez (2016) [64]
1.4) Stopped Vehicles		
Stopped vehicles	Portuguese highway traffic (24/7)	Monteiro et al. (2008) [1-3]
Stopped vehicles	Portuguese highway traffic (24/7)	Monteiro (2009) [4]
1.5) Congestion Detection		
	1-Cars	
Congestion detection	Aerial videos	Lin et al. (2009) [65]
Free-flow/Congestion detection	Urban traffic (India)	Muniruzzaman et al. (2016) [66]
	2-Motorcycles (Motorbikes)	
Helmet detection	Public roads (Brazil)	Silva et al. (2013) [67]
Helmet detection	Naresuan University Campus (Thailand)	Waranusast et al. (2013) [68]
Helmet detection	Indian Institute of Technology Hyderabad (India)	Dahiya et al. (2016) [69]
	3-Pedestrians	
Pedestrian abnormal behavior	Public roads (China)	Jiang et al. (2015) [70]
Airport Surveillance	**1-Airplanes**	
AVITRACK Project (European)	Airports apron	Blauensteiner and Kampel (2004) [38]
AVITRACK Project (European)	Airports apron	Aguilera (2005) [71]
AVITRACK Project (European)	Airports apron	Thirde (2006) [72]
AVITRACK Project (European)	Airports apron	Aguilera (2006) [73]
	2-Ground Vehicles (Fueling vehicles/ Baggage cars)	
AVITRACK Project (European)	Airports apron	Blauensteiner and Kampel (2004) [38]
AVITRACK Project (European)	Airports apron	Aguilera (2005) [71]
AVITRACK Project (European)	Airports apron	Thirde (2006) [72]
AVITRACK Project (European)	Airports apron	Aguilera (2006) [73]
	3-People (Workers)	
AVITRACK Project (European)	Airports apron	Blauensteiner and Kampel (2004) [38]
AVITRACK Project (European)	Airports apron	Aguilera (2005) [71]

(Continued)

TABLE 4.1 *(Continued)*

Visual Surveillance of Human Activities

Subcategories—Aims	Objects of Interest—Scenes	Authors—Dates
AVITRACK Project (European)	Airports apron	Thirde (2006) [72]
AVITRACK Project (European)	Airports apron	Aguilera (2006) [73]
Maritime Surveillance	**1-Cargos**	
	Ocean at Miami (US)	Culibrk et al. (2006) [74]
	Harbor scenes (Ireland)	Zhang et al. (2012) [75]
	2-Boats	
Stationary camera	Miami canals (US)	Socek et al. (2005) [76]
Dock inspecting event	Harbor scenes (China)	Ju et al. (2008) [77]
Different kinds of targets	Bai Chay Beach (Vietnam)	Tran and Le (2016) [78]
Salient events (coastal environments)	Nantucket Island (US)	Cullen et al. (2012) [79]
Salient events (coastal environments)	Nantucket Island (US)	Cullen (2012) [80]
	3-Sailboats	
Sailboat detection	UCSD background subtraction dataset	Sobral et al. (2015) [81]
	4-Ships	
	Italy	Bloisi et al. (2014) [39]
	Fixed ship-borne camera (China) (IR)	Liu et al. (2014) [40]
Different kinds of targets	Ocean (South Africa)	Szpak and Tapamo (2011) [82]
Cage aquaculture	Ocean (Taiwan)	Hu et al. (2011) [83]
	Ocean (Korea)	Arshad et al. (2010) [84]
	Ocean (Korea)	Arshad et al. (2011) [85]
	Ocean (Korea)	Arshad et al. (2014) [86]
	Ocean (Korea)	Saghafi et al. (2012) [87]
Overloaded ship identification	Ocean (China)	Xie et al. (2012) [88]
Ship-bridge collision	Wuhan Yangtze River (China)	Zheng et al. (2013) [89]
	Wuhan Yangtze River (China)	Mei et al. (2017) [90]
	5-Motor Vehicles	
Salient events (coastal environments)	Nantucket Island (US)	Cullen et al. (2012) [79]
Salient events (coastal environments)	Nantucket Island (US)	Cullen (2012) [80]
	6-People (Shoreline)	
Salient events (coastal environments)	Nantucket Island (US)	Cullen et al. (2012) [79]
Salient events (coastal environments)	Nantucket Island (US)	Cullen (2012) [80]
	7-Floating Objects	
Detection of drifting mines	Floating test targets (IR)	Borghgraef et al. (2010) [91]
Store Surveillance	**People**	
Apparel retail store	Panoramic camera	Leykin and Tuceryan (2005) [17, 18]
Apparel retail store	Panoramic camera	Leykin and Tuceryan (2007) [16]
Retail store statistics	Top-view camera	Avinash et al. (2012) [19]

activities by indicating the context, the targets of interest, and the publication with the corresponding date and authors to see the progress. Within this section, we survey the different scenarios and background subtraction methods used by applications for maritime surveillance, airport surveillance, and road surveillance. The idea in all these applications is the automatic detection of static or moving objects according to the following:

- **Static foreground detection (SFD):** Publications discussing the detection of static objects applied to abandoned objects can be found in Porikli et al. [32] and in a more detailed manner specifically in the background model in Cuevas et al. [33].

- **Moving foreground detection (MFD):** Settings where the detection of moving objects is necessary include roads [34–37], airports [38], and maritime surveillance [39, 40]. People, boats, luggage, or airplanes are the most common objects of interest to detect as well as their statistical information. Another important usage for surveillance systems is obtaining information about the behavior of customers within a store [17–19, 41, 42].

4.3.1 Road Surveillance

Road surveillance predominantly requires vehicle detection for traffic controls such as counting vehicles, stopped vehicles, and congestion detection. Road surveillance videos present specific features such as the nature of the environment, type of objects of interest, and the location of the video camera as follows:

- **Camera location:** The videos in traffic surveillance can be classified in three ways: (1) those taken by a fixed camera that is the most common case, in some cases mounted at high altitude (Gao et al. [92]) or not, (2) aerial videos (Reilly [45], Teutsch et al. [35], Eltantawy [93–97]), and very high-resolution satellite videos (Kopsiaftis and Karantzalos [98]).

- **Camera quality:** The video cameras used in closed-circuit television (CCTV) can be low-quality or high-quality (HD cameras). Low-quality cameras usually work with one or two frames/second and 100,000 pixels of resolution per frame in order to reduce cost and disk storage. The amount of information generated by a low-quality camera is approximately 1 GB to 10 GB per day.

- **Environment:** Traffic scenes present different challenges, for example, on roads the movement of tree foliage caused by wind near highways, shadows and changes in lighting and urban traffic, changes in lighting, obstructions, or light reflections.

- **Foreground objects:** The definition of foreground objects in terms of road surveillance refers to the different roles and appearance that users have (behavior, color, or shape). Foreground objects can be pedestrians crossing a street or avenue, cyclists, or users driving any type of vehicle (motorcycles, compact cars, trucks, etc.).

The characteristics described above lead to specific needs and challenges that, according to Song and Tai [99] and Hao et al. [47], can be classified as follows:

1. **Background pixel values:** Background pixels appear more frequently in the scene in a certain position. So, the background intensity can be determined by analyzing the intensity histogram. However, sensing variation and noise may result in

erroneous estimations and cause a foreground object to have the maximum intensity frequency in the histogram.

2. **Challenges due to the camera:** A video camera placed in a high position can be blown into a different position by the wind [100, 101]. In the case of aerial videos, detection has particular challenges due to high object distance, simultaneous object and camera motion, shadows, or weak contrast [35, 45, 93–97]. With satellite videos, the small sizes of the objects and weak contrast are the main challenges [98]. In the case of low-quality cameras, the video is low-quality and noisy, with compression artifacts a low frame rate [63].

3. **Challenges due to the environment:** The background of the scene changes drastically over time due to illumination changes and shadows from the objects that are part of the background. However, a traffic surveillance system must be able to tolerate all these variations regardless of the time of day and weather conditions. Frequently, the shadows are not static due to movement of the trees caused by the wind.

4. **Challenges due to foreground objects:** Camouflage problems can be due to similarities between the colors of the road or shadows and the objects of interest. This causes an object to be detected as background or vice versa. Vehicles can temporarily stop their movement due to traffic signals or traffic lights, causing a severe degradation in the background estimation quality by increasing the weight of the foreground Gaussian distributions. The challenge is greater when the streets or avenues are occupied by a large number of vehicles due to possible overlapping detections. False detections during the night due to vehicle headlights were studied in Li et al. [57].

5. **Challenges in the implementation:** A high processing time will not allow the algorithms to be implemented in real-time as is required in most applications.

Table 4.2 shows an overview of the different publications in the field of traffic surveillance with information about the background model, background maintenance, foreground detection, color space, and strategies used by the authors. The authors used a unimodal model or a multimodal model based on where the camera was placed. For example, if the camera mainly films the road, the most frequently used models are unimodal models such as the median, the histogram, and the single Gaussian. If the camera is in a dynamic environment with waving trees, multimodal models such as mixture of Gaussian (MoG) models are most likely to be used.

For the color space, the authors often used the RGB color space but intensity and YCrCb are also employed to be more robust against illumination changes. Additional strategies mostly focus on shadow detection because it is the most common challenge in this kind of application.

4.3.2 Airport Surveillance

Airport visual surveillance mainly concerns the area where airplanes are parked and maintained by specialized ground vehicles, such as fueling vehicles and baggage cars, as well as tracking of individuals, such as workers. Reasons for airport visual surveillance include: (1) an airport's apron is a security-relevant area, (2) it helps to improve transit time, i.e. the time the aircraft is parked on the apron, and (3) it helps to minimize costs for the company operating the airport, as personnel can be deployed more efficiently, and to

TABLE 4.2

Visual Surveillance of Roads: An Overview

Human Activities	Type	Background Model	Background Maintenance	Foreground Detection	Color Space	Strategies
Traffic surveillance	**1) Road Surveillance**					
	1.1) Road/ Highway traffic					
	Zheng et al. (2006) [34]	Histogram mode	Blind maintenance	Difference	RGB	Aggregation for small range
	Batista et al. (2006) [6]	Average	Running average	Minimum	RGB	Double backgrounds
	Monteiro et al. (2008) [1]	Median	Sliding median	Minimum	RGB	Double backgrounds-shadow/ highlight removal
	Monteiro et al. (2008) [2]	Median	Sliding median	Minimum	RGB	Double backgrounds-shadow/ highlight removal
	Monteiro et al. (2008) [3]	Codebook [106]	Idem codebook	Idem codebook	RGB	Shadow/ highlight detection [107]
	Monteiro (2009) [4]	Sliding median		Minimum	RGB	Double backgrounds-shadow/ highlight removal
	Hao et al. (2013) [102]	KDE [101]	Idem KDE	Idem KDE	Joint domain-range features [101]	Foreground model
	Ling et al. (2014) [36]	Dual-layer approach [36]				
	Sawalakhe and Metkar (2014) [60]	Spatio-temporal BS/ FD (ST-BSFD)	—	AND	RGB	—
	Huang and Chen (2013) [103]	Cerebellar-model-articulation-controller (CMAC) [103]	—	Threshold	YCrCb	Block
	Chen and Huang (2014) [104]	PCA-based RBF network [104]	Selective maintenance	Euclidean distance	YCrCb	Block
	Lee et al. (2015) [49]	GDSM with running average [49]	Selective maintenance	AND	RGB	Shadow detection [108]
	Aqel et al. (2015) [52]	SG on intensity transition	Idem SG	Idem SG	RGB	—

(Continued)

TABLE 4.2 *(Continued)*

Visual Surveillance of Roads: An Overview

Human Activities	Type	Background Model	Background Maintenance	Foreground Detection	Color Space	Strategies
Traffic surveillance *(continued)*	Aqel et al. (2016) [53]	SG on intensity transition	Idem SG	Idem SG	RGB	Shadow detection
	Wang et al. (2016) [105]	Median				
	Dey and Praveen (2016) [61]	GDSM with background subtraction [5]	Yes	AND	RGB	Post-processing
	Hadiuzzaman et al. (2017) [58]	Median	Yes	Idem median	Intensity	Shadow detection
	1.2) Urban Traffic					
	Conventional camera					
	Hwang et al. (2009) [43]	MoG-ALR [43]	Adaptive learning Rate	Idem MoG	RGB	—
	Intachak and Kaewapichai (2011)[107]	Mean (clean images)	Selective maintenance	Idem mean	RGB	Illumination adjustment
	Milla et al. (2013) [48]	Σ - Δ filter	Σ - Δ filter	Σ - Δ filter	Intensity	Short-term/ Long-term backgrounds
	Toropov et al. (2015) [63]	MoG [111]	Idem MoG	Idem MoG	Color	Brightness adjustment
	Zhang et al. (2016) [109]	GMMCM [109]	Idem MoG [111]	Confidence period	Intensity	Classification of traffic density
	Highly mounted camera					
	Gao et al. (2014) [92]	SG [112]	Idem SG	Idem SG	YCrCb	Shadow detection [117]
	Quesada and Rodriguez (2016)[61]	incPCP [113]	Idem incPCP	Idem incPCP	Intensity	—
	Headlight removal					
	Li et al. (2016) [57]	GMM [111]	Idem GMM	Idem GMM	RGB	Headlight/ shadow removal
	Intersection					
	Ding et al. (2012) [46]	CPS-based GMM [46]	Idem GMM	Idem GMM	RGB	Cyber physical system
	Ding et al. (2012) [46]	CPS-based FGD [114]	Idem FGD	Idem FGD	RGB	Cyber physical system
	Alldieck (2015) [51]	Zivkovic-Heijden GMM [115]	Idem GMM	Idem GMM	RGB/IR	Multimodal cameras

(Continued)

TABLE 4.2 *(Continued)*

Visual Surveillance of Roads: An Overview

Human Activities	Type	Background Model	Background Maintenance	Foreground Detection	Color Space	Strategies
Traffic surveillance *(continued)*	Mendoca et Oliveira (2015)[126]	Context supported road information (CRON)[126]	Selective maintenance	Idem AM	RGB	
	Li et al. (2016) [62]	incPCP [116]	Idem incPCP	Idem incPCP	RGB	—
	Obstacle detection					
	Lan et al. (2015) [110]	SUOG [110]	Selective maintenance	Idem GMM	RGB	Obstacle detection model
	2) Vehicle Counting					
	Unzueta et al. (2012)[37]	Multicue approach [37]				
	Virginas-Tar et al. (2014)[7]	MoG-EM [118]	Idem MoG-EM	Idem MoG-EM		Shadow detection [107]
	3) Vehicle Detection					
	3.1) Conventional video					
	Wang and Song (2011)[44]	GMM with spatial correlation method [44]	Idem GMM	Idem GMM	HSV	Spatial correlation method
	Hadi et al. (2014)[50]	Histogram mode	Blind maintenance	Absolute difference	RGB	Morphological processing
	Hadi et al. (2017) [5]	Histogram mode	Blind maintenance	Absolute difference	RGB	Morphological processing
	3.2) Aerial video					
	Lin et al. (2009) [65]	Two CFD	—	—	RGB	Translation
	Reilly (2012) [45]	Median	Idem median	Idem median	Intensity	—
	Teutsch et al. (2014)[35]	Independent motion detection [35]				
	3.3) Satellite video					
	Kopsiaftis and Karantzalos (2015)[98]	Mean	Idem mean	Idem mean	Intensity	—
	Yang et al. (2016)[119]	Local saliency-based background model based on ViBe (LS-ViBe) [119]	Idem ViBe	Idem ViBe	Intensity	—

(Continued)

TABLE 4.2 *(Continued)*

Visual Surveillance of Roads: An Overview

Human Activities	Type	Background Model	Background Maintenance	Foreground Detection	Color Space	Strategies
Traffic surveillance *(continued)*	**4) Illegally Parked Vehicles**					
	Lee et al. (2007) [8]	Median	Selective maintenance	Difference	RGB	Morphological processing
	Zhao et al. (2013) [9]	Average	Running average	Difference	HIS	Morphological processing-salient object detection
	Saker et al. (2015) [10]	GMM-PUC [120]	Idem GMM	Idem GMM	RGB	Detection of stationary object
	Chunyang et al. (2015)[11]	MoG [111]	Idem MoG	Idem MoG	RGB	Morphological processing
	Wahyono et al. (2015) [12]	Running average	Selective maintenance	Difference	RGB	Dual background models
	5) Vacant Parking Area					
	Postigo et al. (2015)[13]	MoG-EM [118]	Idem MoG-EM	Idem MoG-EM	RGB	Transience map
	Neuhausen (2015) [14]	SG	Selective maintenance	Choquet integral [121]	YCrCb-ULBP [122]	Adaptive weight on illumination normalization
	6) Motorcycle (Motorbike) Detection					
	Silva et al. (2013) [67]	Zivkovic-Heijden GMM [115]	Idem GMM	Idem GMM	Intensity	—
	Waranusast et al. (2013)[68]	Zivkovic-Heijden GMM [115]	Idem GMM	Idem GMM	RGB	Morphological operators
	Dahiya et al. (2016) [69]	Zivkovic-Heijden GMM [115]	Idem GMM	Idem GMM	Intensity	Detection bike riders

minimize latencies for the passengers, as less time is needed for accomplishing ground services. Practically, airport surveillance videos present their own characteristics. Weather and light changes are very challenging problems. Ground support vehicles (e.g., a baggage car) may change their shape to an extended degree during the tracking period. Strictly rigid motion and object models may not work on tracking those vehicles. Vehicles may build blobs, either with the aircraft or with other vehicles, for a longer period of time, e.g., a fueling vehicle during the refueling process.

A global description of different publications in the area of airport surveillance is shown in Table 4.3, providing information about the techniques used, characteristics of the background model, foreground detection, and the color space used. It can be noted that the authors

TABLE 4.3

Visual Surveillance of Airport and Maritime Zones: An Overview (Part I)

Human Activities	Type	Background Model	Background Maintenance	Foreground Detection	Color Space	Strategies
Airport surveillance	Blauensteiner and Kampel (2004)[38]	Median	Idem Median	Idem Median	RGB	—
	Aguilera et al. (2005) [71]	Single Gaussian	Idem SG	Idem SG	RGB	—
	Thirde et al. (2006) [72]	Single Gaussian	Idem SG	Idem SG	RGB	—
	Aguilera et al. (2006) [73]	Single Gaussian	Idem SG	Idem SG	RGB	—
Maritime surveillance	**1) Fluvial Canals Environment**					
	Socek et al. (2005) [76]	Two CFD	—	Bayesian decision [17]	RGB	Color segmentation
	Bloisi et al. (2014) [39]	IMBS [124]				
	2) River Environment					
	Zheng et al. (2013) [89]	LBP histogram [125]	LBP histogram[125]	LBP histogram [125]	RGB	—
	Mei et al. (2017) [90]	EAdaDGS [90]	Improved mechanism	Idem AdaDGS [127]	RGB	Multiresolution
	3) Open-Sea Environment					
	Culibrk et al. (2006) [74]	MoG-GNN [111]	Idem MoG	Idem MoG	Intensity	—
	Zhang et al. (2009) [123]	Median				
	Borghgraef et al. (2010) [91]	Zivkovic-Heijden GMM [115]	Idem Zivkovic-Heijden GMM	Idem Zivkovic-Heijden GMM	IR	—
	Arshad et al. (2010) [84]	—	—	—	RGB	Morphological processing
	Arshad et al. (2011) [85]	—	—	—	RGB	Morphological processing
	Szpak and Tapamo (2011)[82]	Single Gaussian [112]	Idem SG	Idem SG	Intensity	Spatial smoothness
	Hu et al. (2011) [83]	Modified AM [83]	Idem modified AM	Idem modified AM	RGB	Fast 4-Connected component Labeling
	Saghafi et al. (2012)[84]	Modified ViBe [87]	Modified ViBe	Modified ViBe	RGB	backwash cancellation algorithm

(Continued)

TABLE 4.3 *(Continued)*

Visual Surveillance of Airport and Maritime Zones: An Overview (Part I)

Human Activities	Type	Background Model	Background Maintenance	Foreground Detection	Color Space	Strategies
Maritime surveillance *(continued)*	Xie et al. (2012) [88]	Three CFD	Selective maintenance	AND	Intensity	
	Zhang et al. (2012) [75]	PCA135/ GMM [111]	Idem PCA[135]/ GMM[110]	Idem PCA[128]/ GMM[111]	Intensity	—
	Arshad et al. (2014) [86]	—	—	—	RGB	Morphological processing
	Liu et al. (2014) [40]	Modified histogram [40]	Idem histogram	Idem histogram	IR	Adaptive row mean filter
	Sobral et al. (2015) [81]	Double constrained RPCA [81]	Idem RPCA	Idem RPCA	RGB	Saliency detection
	Tran and Le (2016) [78]	ViBe [126]	Idem ViBe	Idem ViBe	RGB	Saliency detection

[38, 71–73] made use of the single Gaussian or the median because the paved road corresponds to a unimodal background in video scenes. In all publications, the color space used was RGB.

4.3.3 Maritime Surveillance

The most common tasks in maritime surveillance are the counting, recognition, or tracking of boats in rivers, fluvial channels, or open sea, either in the visible spectrum [39, 123] or infrared (IR) [40]. Most of the time this involves detection of and counting boats for traffic control in harbors. A hybrid approach to improve accuracy for foreground detection in the fluvial channels of Miami was proposed by Socek et al. [76]. The hybrid approach combines Bayesian background subtraction with a color segmentation technique. For the characterization of multimodal backgrounds caused by the water, independent multimodal background subtraction (IMBS [124]) was used by Bloisi et al. [39], which is a pixel-level, nonpredictive, and nonrecursive model designed to deal with irregular and high-frequency noise. To model background environments in the open ocean, Culibrk et al. [74] combined the GMM [111] with the general regression neural network (GRNN).

In another work of interest, Liu et al. [40] applied modified histogram models to IR videos. For the detection of sailboats, Sobral et al. [81] used as a basis the saliency detection for the development of a RPCA with double constraints. Borghgraef et al. [91] used the Zivkovic-Heijden GMM [115], which is an improved MoG model for the detection of strange floating objects. To detect and track boats, Zhang et al. [123] applied the model of the median. Szpak and Tapamo et al. [82] based their experiments on the single Gaussian to detect and track different types of vessels such as sailboats, tankers, Jet Skis, and patrol boats. Tran and Le et al. [78] concluded that the ViBe algorithm is a better choice than the original MoG to detect different types of boats due to the great variety of shapes, structures, sizes, and textures that can present as fishing boats, cargo ships, canoes, or cruise ships. Hu et al. [83] made use of the approximate median based on AM [129] with a wave ripple removal for automatic detection and tracking of intruder ships in aquaculture cages. A global description in the field of maritime surveillance is shown in Table 4.3 with information regarding the techniques used by the authors, color spaces, background modeling and updating, as well as the detection of the foreground.

Most authors prefer the use of multimodal background models due to the dynamic behavior of water in a real environment. In daytime conditions, the RGB color space is the most used whereas for night conditions, the use of IR is preferred. Additional strategies such as saliency detection and morphological processing are used to solve false-positive problems caused by the continuous movement of water.

4.3.4 Surveillance on the Coasts

The background subtraction model developed by Jodoin et al. [130] was used by Cullen et al. [79, 80] in the detection of events of interest in coastal environments such as the erosion of land, wildlife, and human interaction with the environment, among others, which may be of great interest for organizations. For example, for the protection of marine mammals, biologists are interested to know if humans get too close to these animals. The United States Fish and Wildlife Service is interested in obtaining statistics about how many people and cars are present on the beaches each day to determine if the delicate sand dunes are affected. Cullen et al. [79, 80] detected the presence of motorized vehicles, people, and ships near the coasts during some tests. Table 4.4 shows some publications related to coastal surveillance with information about the algorithms used by the authors, and information about the background model, color space, and foreground detection.

TABLE 4.4

Visual Surveillance of Airport and Maritime Zones: An Overview (Part II)

Human Activities	Type	Background Model	Background Maintenance	Foreground Detection	Color Space	Strategies
Store surveillance	Leykin and Tuceryan (2007)[18]	Codebook [106]	Idem codebook	Idem Codebook	RGB	—
	Leykin and Tuceryan (2005)[19]	Codebook [106]	Idem codebook	Idem Codebook	RGB	—
	Leykin and Tuceryan (2005)[20]	Codebook [106]	Idem codebook	Idem Codebook	RGB	—
	Avinash et al. (2012) [19]	Single Gaussian [112]	Idem SG	Idem SG	RGB	—
	Zhou et al. (2017) [42]	MoG [140]	Idem MoG	Idem MoG	RGB	—
Coastal surveillance	Cullen et al. (2012) [79]	Behavior subtraction [130]	Idem BS	Idem BS	RGB	—
	Cullen (2012) [80]	Behavior subtraction [130]	Idem BS	Idem BS	RGB	—
Swimming pools surveillance	**1) Online Videos**					
	1.1) Top-view videos					
	Eng et al. (2003) [131]	Block-based median [131]	Sliding mean	Block-based difference	CIELa*b*	Partial occlusion handling

(Continued)

TABLE 4.4 *(Continued)*

Visual Surveillance of Airport and Maritime Zones: An Overview (Part II)

Human Activities	Type	Background Model	Background Maintenance	Foreground Detection	Color Space	Strategies
Swimming pools surveillance *(continued)*	Eng et al. (2004) [132]	Region-based single multivariate Gaussian[139]	Idem SG	Region-based SG	CIELa*b*	Handling specular reflection
	Chan (2011) [133]	SG/optical ow [133]	Idem SG	Designed distance	RGB	—
	Chan (2014) [134]	MoG/optical flow [134]	Idem MoG	Idem MoG	RGB	—
	1.2) Underwater videos					
	Fei et al. (2009) [135]	MoG	Idem MoG			Shadow removal
	Lei and Zhao (2010) [136]	Kalman filter [141]	Idem Kalman filter	Idem Kalman filter	RGB	
	Zhang et al. (2015) [137]	—	—	—	RGB	Interframe-based denoising
	2) Archived Videos					
	Sha et al. (2014) [138]	—	—	—	RGB	—
	3) Private Swimming Pools					
	Peixoto et al. (2012) [139]	Mean/two CFD [139]	Selective maintenance	Mean distance/Two CFD Distance	HSV	

4.3.5 Surveillance in Stores

For marketing researchers in industrial or academic fields, decision-making is crucial, so they need computer vision tools to know how customers interact with the environment, something that traditional sensors cannot offer. Statistical parameters for this type of application are customer dwell time and interaction with products of interest. Leykin and Tuceryan [16–18], and Avinash et al. [19] obtained statistical information about shopper groups. Table 4.4 shows the different techniques used by researchers, color space, and foreground detection, as well as information and updating of the background model. Because the scenes correspond to indoors, the authors used both unimodal and multimodal models and RGB color space.

4.4 Solved and Unsolved Challenges, and Possible Solutions

As has been mentioned, each application has its own characteristics in terms of objects of interest and the environment, representing a different challenge in background modeling. Additionally, the publications in the scientific literature for the different applications can

TABLE 4.5

Solved and Unsolved Challenges in Visual Surveillance of Human Activities

Applications	Scenes	Challenges	Solved/ Unsolved
1) Intelligent Visual Surveillance of Human Activities			
1.1) Traffic Surveillance	Outdoor scenes	Multimodal backgrounds	Partially solved
		Illumination changes	Partially solved
		Camera jitter	Partially solved
1.2) Airport Surveillance	Outdoor scenes	Illumination changes	Partially solved
		Camera jitter	Partially solved
		Illumination changes	Partially solved
1.3) Maritime Surveillance	Outdoor scenes	Multimodal backgrounds	Partially solved
		Illumination changes	Partially solved
		Camera jitter	Partially solved
1.4) Store Surveillance	Indoor scenes	Multimodal backgrounds	Solved
1.5) Coastal Surveillance	Outdoor scenes	Multimodal backgrounds	Partially solved
1.6) Swimming Pool Surveillance	Outdoor scenes	Multimodal backgrounds	Solved

be outdated, which represents another challenge when working with certain scenarios. Table 4.5 shows the different pending challenges to be solved, among which are illumination changes, camera jitter, and multimodal backgrounds in outdoor environments. These challenges are classified as solved and unsolved, and according to the application. Future research can focus on these kinds of problems. Possible recent alternatives for the characterization and background subtraction are based on RPCA [142–146] and deep learning [147–150]. In addition, for complex scenarios such as maritime environments, background initialization methods [151–155] as well as deep learned features [156–159] should be designed robustly.

4.5 Conclusion

In this chapter, we have surveyed the main visual surveillance applications of human activities where background subtraction is employed for the detection of moving or static objects. Subsequently, a review of the different challenges was made according to the type of application for the characterization of the scene background and the detection of moving objects. Foreground detection is essential for later tasks such as recognition and behavior analysis. Therefore, it is desirable that the foreground mask be as accurate and fast as possible for real-time applications. Varona et al. [160] carried out a rigorous study of the influence of background subtraction on the subsequent steps. The different critical elements are the types of environments, types of foreground objects, and the location of the video cameras. Due to diverse environments, the background model must deal with different challenges. In addition, the foreground objects have different characteristics depending on the type of application. Therefore, it is ideal to design a background model for each environment and thus select the most appropriate for the corresponding challenges.

For applications in controlled environments such as indoor scenes, the basic models are sufficient for motion detection. However, for challenges such as sleeping/beginning foreground objects and illumination changes in traffic surveillance, the use of statistical models becomes necessary. In addition, background subtraction can be also used in applications in which cameras are slowly moving [161–163]. For example, Taneja et al. [164] proposed to model dynamic scenes recorded with freely moving cameras.

References

1. Monteiro, G., Ribeiro, M., Marcos, J., & Batista, J. (2008). Robust segmentation for outdoor traffic surveillance. In IEEE ICIP 2008.
2. Monteiro, G., Marcos, J., Ribeiro, M., & Batista, J. (2008, June 25-27). Robust segmentation process to detect incidents on highways. In Image Analysis and Recognition: 5th International Conference Proceedings.
3. Monteiro, G., Marcos, J., & Batista, J. (2008). Stopped vehicle detection system for outdoor traffic surveillance. In RecPad 2008: 14th Portuguese Conference on Pattern Recognition, Coimbra.
4. Monteiro, G. (2009). *Traffic video surveillance for automatic incident detection on highways* (Master thesis). University of Coimbra.
5. Hadi, R., George, L., & Mohammed, M. (2017). A computationally economic novel approach for real-time moving multi-vehicle detection and tracking toward efficient traffic surveillance. *Arabian Journal for Science and Engineering, 42*(2), 817–831
6. Batista, J., Peixoto, P., Fernandes, C., & Ribeiro, M. (2006). A dual-stage robust vehicle detection and tracking for real-time traffic monitoring. *IEEE ITSC 2006.*
7. Virginas-Tar, A., Baba, M., Gui, V., Pescaru, D., & Jian, I. (2014). Vehicle counting and classification for traffic surveillance using wireless video sensor networks. In Telecommunications Forum Telfor, TELFOR 2014, pp. 1019–1022.
8. Lee, J., Ryoo, M., Riley, M., & Aggarwal, J. (2007). Real-time detection of illegally parked vehicles using 1-D transformation. In *IEEE AVSS 2007.*
9. Zhao, X., Cheng, X., & Li, X. (2013). Illegal vehicle parking detection based on online learning. In World Congress on Multimedia and Computer Science.
10. Saker, M., Weihua, C., & Song, M. (2015). Detection and recognition of illegally parked vehicle based on adaptive Gaussian mixture model and seed fill algorithm. *Journal of Information and Communication Convergence Engineering,* 197–204.
11. Chunyang, M., Xing, M. W., & Panpan, Z. (2015). Smart detection of vehicle in illegal parking area by fusing of multi-features. In 9th International Conference on Next Generation Mobile Applications, Services and Technologies, pp. 388–392.
12. Wahyono, Filonenko, A., & Jo, K. (2015). Illegally parked vehicle detection using adaptive dual background model. In IECON 2015, pp. 2225–2228.
13. Postigo, C., Torres, J., & Menendez, J. (2015). Vacant parking area estimation through background subtraction and transience map analysis. *IET ITS, 9*(9), 835–841.
14. Neuhausen, M. (2015). Video-based parking space detection: Localization of vehicles and development of an infrastructure for a routing system. In *Institut fur Neuroinformatik.*
15. Cho, J., Park, J., Baek, U., Hwang, D., Choi, S., Kim, S., & Kim, K. (2016). Automatic parking system using background subtraction with CCTV environment. *ICCAS 2016*, 1649–1652.

16. Leykin, A., & Tuceryan, M. (2007). Detecting shopper groups in video sequences. In IEEE AVSBS 2007.
17. Leykin, A., & Tuceryan, M. (2005). A vision system for automated customer tracking for marketing analysis: Low level feature extraction. In HAREM 2005.
18. Leykin, A., & Tuceryan, M. (2005). Tracking and activity analysis in retail environments. In TR 620 Computer Science Department, Indiana University.
19. Avinash, N., Kumar, M. S., & Sagar, S. (2012). Automated video surveillance for retail store statistics generation. In ICSIP 2012, pp. 585–596.
20. Tamas, B. (2016). Detecting and analyzing rowing motion in videos. In BME Scientific Student Conference.
21. Bastos, R. (2015). *Monitoring of human sport activities at sea* (Master thesis). *Portugal.*
22. Bouwmans, T., Baf, F. E., & Vachon, B. (2010, January). Statistical background modeling for foreground detection: A survey. *Handbook of Pattern Recognition and Computer Vision, World Scientific Publishing, 4*(2), 181–199.
23. Bouwmans, T. (2012, March). Background subtraction for visual surveillance: A fuzzy approach. In *Handbook on Soft Computing for Video Surveillance* (pp. 103–139). Taylor and Francis Group.
24. Bouwmans, T., Baf, F. E., & Vachon, B. (2008). Background modeling using mixture of Gaussians for foreground detection—A survey. *RPCS 2008, 1*(3), 219–237.
25. Bouwmans, T. (2009). Subspace learning for background modeling: A survey. *RPCS 2009, 2*(3), 223–234.
26. Bouwmans, T. (2011). Recent advanced statistical background modeling for foreground detection: A systematic survey. *RPCS 2011, 4*(3), 147–176.
27. Bouwmans, T., & Zahzah, E. (2014). Robust PCA via principal component pursuit: A review for a comparative evaluation in video surveillance. *CVIU 2014, 122,* 22–34.
28. Bouwmans, T. (2014). Traditional and recent approaches in background modeling for foreground detection: An overview. *Computer Science Review, 11,* 31–66.
29. Bouwmans, T., Sobral, A., Javed, S., Jung, S., & Zahzah, E. (2017). Decomposition into low-rank plus additive matrices for background/foreground separation: A review for a comparative evaluation with a large-scale dataset. *Computer Science Review, 23,* 1–71.
30. Bouwmans, T., Silva, C., Marghes, C., Zitouni, M., Bhaskar, H., & Frelicot, C. (2018). On the role and the importance of features for background modeling and foreground detection, *Computer Science Review, 28,* 26–91.
31. Maddalena, L., & Petrosino, A. (2018). Background subtraction for moving object detection in rgb-d data: A survey. *MDPI Journal of Imaging, 4,* 71.
32. Porikli, F., Ivanov, Y., & Haga, T. (2008). Robust abandoned object detection using dual foregrounds. In EURASIP *Journal on Advances in Signal Processing,* p. 30.
33. Cuevas, C., Martinez, R., & Garcia, N. (2016). Detection of stationary foreground objects: A survey. *Computer Vision and Image Understanding, 152.*
34. Zheng, J., Wang, Y. H., Nihan, N., & Hallenbeck, M. (2006). Extracting roadway background image: A mode based approach. In *Transportation Research Report,* pp. 82–88.
35. Teutsch, M., Krueger, W., & Beyerer, J. (2014, August). Evaluation of object segmentation to improve moving vehicle detection in aerial videos. In IEEE AVSS 2014.
36. Ling, Q., Yan, J., Li, F., & Zhang, Y. (2014). A background modeling and foreground segmentation approach based on the feedback of moving objects in traffic surveillance systems. *Neurocomputing, 133,* 32–45.
37. Unzueta, L., Nieto, M., Cortes, A., Barandiaran, J., Otaegui, O., & Sanchez, P. (2012, June). Adaptive multi-cue background subtraction for robust vehicle counting and classification. *IEEE T-ITS, 13*(2), 527–540.
38. Blauensteiner, P., & Kampel, M. (2004). Visual surveillance of an airport's apron—an overview of the AVITRACK project. In AAPR 2004, pp. 213–220.
39. Bloisi, D., Pennisi, A., & Iocchi, L. (2014, July). Background modeling in the maritime domain. *Machine Vision and Applications, 25*(5), 1257–1269.

40. Liu, Z., Zhou, F., Chen, X., Bai, X., & Sun, C. (2014). Iterative infrared ship target segmentation based on multiple features. *Pattern Recognition, 47*(9), 2839–2852.

41. Lee, S., Kim, N., Paek, I., Hayes, M., & Paik, J. (2013, January). Moving object detection using unstable camera for consumer surveillance systems. In ICCE 2013, pp. 145–146.

42. Zhou, Z., Shangguan, L., & Liu, Y. (2017). Design and implementation of an rfid-based customer shopping behavior mining system. In IEEE/ACM Transactions on Networking.

43. Hwang, P., Eom, K., Jung, J., & Kim, M. (2009). A statistical approach to robust background subtraction for urban traffic video. In International Workshop on Computer Science and Engineering, pp. 177–181.

44. Wang, C., & Song, Z. (2011). Vehicle detection based on spatial-temporal connection background subtraction. In IEEE ICIA, pp. 320–323.

45. Reilly, V. (2012). *Detecting, tracking, and recognizing activities in aerial video* (PhD thesis). University of Central Florida.

46. Ding, R., Liu, X., Cui, W., & Wang, Y. (2012). Intersection foreground detection based on the cyber-physical systems. In IET ICISCE 2012, pp. 1–7.

47. Hao, J., Li, C., Kim, Z., & Xiong, Z. (2013, March). Spatio-temporal traffic scene modeling for object motion detection. *IEEE T-ITS, 1*(14), 295–302.

48. Milla, J., Toral, S., Vargas, M., & Barrero, F. (2013). Dual-rate background subtraction approach for estimating traffic queue parameters in urban scenes. In IET ITS.

49. Lee, G., Mallipeddi, R., Jang, G., & Lee, M. (2015). A genetic algorithm-based moving object detection for real-time traffic surveillance. *IEEE SPL, 10*(22), 1619–1622.

50. Hadi, R., Sulong, G., & George, L. (2014). An innovative vehicle detection approach based on background subtraction and morphological binary methods. *Life Science Journal*, 230–238.

51. Alldieck, T. (2015). *Information based multimodal background subtraction for traffic monitoring applications* (Master thesis). Aalborg University, Danemark.

52. Aqel, S., Sabri, M., & Aarab, A. (2015). Background modeling algorithm based on transitions intensities. *IRECOS 2015, 10*(4).

53. Aqel, S., Aarab, A., & Sabri, M. (2016). Traffic video surveillance: Background modeling and shadow elimination. *IT4OD 2016*, 1–6.

54. Wang, K., Liu, Y., Gou, C., & Wang, F. (2015). A multi-view learning approach to foreground detection for traffic surveillance applications. In IEEE T-VT.

55. Zhang, Y., Zhao, C., He, J., & Chen, A. (2016). Vehicles detection in complex urban traffic scenes using Gaussian mixture model with confidence measurement. In IET ITS.

56. Hargude, S., & Idate, S. (2016). i-surveillance: Intelligent surveillance system using background subtraction technique. In ICCUBEA 2016, pp. 1–5.

57. Li, Q., Bernal, E., & Loce, R. (2016). Scene-independent feature- and classifier-based vehicle headlight and shadow removal in video sequences. In IEEE WACVW 2016, pp. 1–8.

58. Hadiuzzaman, M., Haque, N., Rahman, F., Hossain, S., Rayeedul, M., Siam, K., & Qiu, T. (2017, April). Pixel-based heterogeneous traffic measurement considering shadow and illumination variation. *Signal, Image and Video Processing*, 1–8.

59. Qu, Z., & Liu, J. (2010, June). Real-time traffic vehicle tracking based on improved MoG background extraction and motion segmentation. ISSCAA 2010, 676–680.

60. Sawalakhe, S., & Metkar, S. (2014). Foreground background traffic scene modeling for object motion detection. In IEEE India Conference, INDICON 2014, pp. 1–6.

61. Dey, J. & Praveen, N. (2016). Moving object detection using genetic algorithm for traffic surveillance. In ICEEOT 2016, pp. 2289–2293.

62. Li, C., Chiang, A., Dobler, G., Wang, Y., Xie, K., Ozbay, K., & Wang, D. (2016). Robust vehicle tracking for urban traffic videos at intersections. In IEEE AVSS 2016.

63. Toropov, E., Gui, L., Zhang, S., Kottur, S., & Moura, J. (2015). Traffic flow from a low-frame rate city camera. In IEEE ICIP 2015.

64. Quesada, J. & Rodriguez, P. (2016). Automatic vehicle counting method based on principal component pursuit background modeling. In IEEE ICIP 2016.

65. Lin, R., Cao, X., Xu, Y., Wu, C., & Qiao, H. (2009). Airborne moving vehicle detection for video surveillance of urban traffic, IEEE IVS 2009, pp. 203–208.
66. Muniruzzaman, S., Haque, N., Rahman, F., Siam, M., Musabbir, R., & Hossain, S. (2016). Deterministic algorithm for traffic detection in free-flow and congestion using video sensor. *Journal of Built Environment, Technology and Engineering, 1,* 111–130.
67. Silva, R., Aires, K., Santos, T., Abdala, K., Veras, R., & Soares, A. (2013). Automatic detection of motorcyclists without helmet. In CLEI 2013, pp. 1–7.
68. Waranusast, R., Bundon, N., & Pattanathaburt, P. (2013). Machine vision techniques for motor-cycle safety helmet detection. In IVCNZ 2013, pp. 35–40.
69. Dahiya, K., Singh, D., & Mohan, C. (2016). Automatic detection of bike-riders without helmet using surveillance videos in real-time. In IJCNN 2016, pp. 3046–3051.
70. Jiang, Q., Li, G., Yu, J., & Li, X. (2015). A model based method for pedestrian abnormal behavior detection in traffic scene. In IEEE ISC2 2015, pp. 1–6.
71. Aguilera, J., Wildenauer, H., & Ferryman, J. (2005). Evaluation of motion segmentation quality for aircraft activity surveillance. In IEEE PETS 2005, pp. 293–300.
72. Thirde, D., Borg, M., & Ferryman, J. (2006). A real-time scene understanding system for airport apron monitoring. In ICVS 2006.
73. Aguilera, J., Thirde, D., Kampel, M., Borg, M., Fernandez, G., & Ferryman, J. (2006). Visual surveillance for airport monitoring applications. In CVW 2006.
74. Culibrk, D., Marques, O., Socek, D., & Furht, B. (2006, February). A neural network approach to Bayesian background modeling for video object segmentation. In VISAPP 2006.
75. Zhang, D., O'Connor, E., & Smeaton, A. (2012). A visual sensing platform for creating a smarter multi-modal marine monitoring network. In MAED 2012, pp. 53–56.
76. Socek, D., Culibrk, D., Marques, O., & Furht, B. (2005). A hybrid color-based foreground object detection method for automated marine surveillance. In ACVIS 2005.
77. Ju, S., Chen, X., & Xu, G. (2008). An improved mixture Gaussian models to detect moving object under real-time complex background. In ICC 2008, pp. 730–734.
78. Tran, T., & Le, T. (2016). Vision based boat detection for maritime surveillance. In ICEIC 2016, pp. 1–4.
79. Cullen, D., Konrad, J., & Little, T. (2012). Detection and summarization of salient events in coastal environments. In IEEE AVSS 2012.
80. Cullen, D. (2012, May). Detecting and summarizing salient events in coastal videos. In TR 2012–06, Boston University.
81. Sobral, A., Bouwmans, T., & Zahzah, E. (2015). Double-constrained RPCA based on saliency maps for foreground detection in automated maritime surveillance. In ISBC 2015 Workshop conjunction with AVSS 2015.
82. Szpak, Z., & Tapamo, J. (2011). Maritime surveillance: Tracking ships inside a dynamic background using a fast level-set. *ESWA, 38*(6), 6669–6680.
83. Hu, W., Yang, C., & Huang, D. (2011, August). Robust real-time ship detection and tracking for visual surveillance of cage aquaculture. *JVCIR, 22*(6), 543–556.
84. Arshad, N., Moon, K., & Kim, J. (2010). Multiple ship detection and tracking using background registration and morphological operations. *International Conference on Signal Processing and Multimedia Communication, 123,* 121–126.
85. Arshad, N., Moon, K., & Kim, J. (2011). An adaptive moving ship detection and tracking based on edge information and morphological operations. In SPIE ICGIP 2011.
86. Arshad, N., Moon, K., & Kim, J. (2014, September). Adaptive real-time ship detection and tracking using morphological operations. *Journal of Information Communication Convergence Engineering, 3*(12), 168–172.
87. Saghafi, M., Javadein, S., Noorhossein, S., & Khalili, H. (2012). Robust ship detection and tracking using modified ViBe and backwash cancellation algorithm. In CIIT 2012.
88. Xie, L., Li, B., Wang, Z., & Yan, Z. (2012, December). Overloaded ship identification based on image processing and Kalman filtering. *Journal of Convergence Information Technology, 7*(22), 425–433.

89. Zheng, Y., Wu, W., Xu, H., Gan, L., & Yu, J. (2013). A novel ship-bridge collision avoidance system based on monocular computer vision, *Research Journal of Applied Sciences, Engineering and Technology*, 4(6), 647–653.
90. Mei, L., Guo, J., Lu, P., Liu, Q., & Teng, F. (2017). Inland ship detection based on dynamic group sparsity. In ICACI 2017, pp. 1–6.
91. Borghgraef, A., Barnich, O., Lapierre, F., Droogenbroeck, M., Philips, W., & Acheroy, M. (2010). An evaluation of pixel-based methods for the detection of floating objects on the sea surface. In EURASIP *Journal on Advances in Signal Processing*.
92. Gao, Z., Wang, W., Xiong, H., Yu, W., Liu, X., Liu, W., & Pei, L. (2014). Traffic surveillance using highly-mounted video camera. In IEEE International Conference on Progress in Informatics and Computing, pp. 336–340.
93. Eltantawy, A., & Shehata, M. (2016). A novel method for segmenting moving objects in aerial imagery using matrix recovery and physical spring model. In ICPR 2016, pp. 3898–3903.
94. Eltantawy, A., & Shehata, M. (2015). Moving object detection from moving platforms using Lagrange multiplier. In IEEE ICIP 2015.
95. Eltantawy, A., & Shehata, M. (2015). UT-MARO: unscented transformation and matrix rank optimization for moving objects detection in aerial imagery. In ISVC 2015, pp. 275–284.
96. Eltantawy, A., & Shehata, M. (2018, May). MARO: Matrix rank optimization for the detection of small-size moving objects from aerial camera platforms. *Signal, Image and Video Processing*, 12(4), 641–649.
97. Eltantawy, A., & Shehata, M. (2018, June). KRMARO: Aerial detection of small-size ground moving objects using kinematic regularization and matrix rank optimization. In IEEE Transactions on Circuits and Systems for Video Technology.
98. Kopsiaftis, G., & Karantzalos, K. (2015). Vehicle detection and traffic density monitoring from very high resolution satellite video data. In IEEE IGARSS 2015, pp. 1881–1884.
99. Song, K., & Tai, J. (2008). Real-time background estimation of traffic imagery using group-based histogram. *Journal of Information Science and Engineering*, 24, 411–423.
100. Sheikh, Y., & Shah, M. (2005, June). Bayesian object detection in dynamic scenes. In IEEE Conference on Computer Vision and Pattern Recognition, CVPR 2005.
101. Sheikh, Y., & Shah, M. (2005, November). Bayesian modeling of dynamic scenes for object detection. *IEEE T-PAMI*, 27(11), 1778–1792.
102. Hao, J., Li, C., Kim, Z., & Xiong, Z. (2013, March). Spatio-temporal traffic scene modeling for object motion detection. *IEEE T-ITS*, 1(14), 295–302.
103. Huang, S., & Chen, B. (2013, December). Highly accurate moving object detection in variable bit rate video-based traffic monitoring systems. *IEEE T-NNLS*, 24(12), 1920–1931.
104. Chen, B., & Huang, S. (2014, April). An advanced moving object detection algorithm for automatic traffic monitoring in real-world limited bandwidth networks. *IEEE Transactions on Multimedia*, 16(3), 837–847.
105. Wang, K., Liu, Y., Gou, C., & Wang, F. (2016, June). A multi-view learning approach to foreground detection for traffic surveillance applications. *IEEE Transactions on Vehicular Technology*, 65(6), 4144–4158.
106. Kim, K., Chalidabhongse, T., Harwood, D., & Davis, L. (2004). Background modeling and subtraction by codebook construction. In IEEE ICIP 2004.
107. Horprasert, T., Harwood, D., & Davis, L. (1999, September). A statistical approach for real-time robust background subtraction and shadow detection. In IEEE International Conference on Computer Vision, FRAME-RATE Workshop.
108. Cucchiara, R., Grana, C., Piccardi, M., Prati, A., & Sirottia, S. (2001). Improving shadow suppression in moving object detection with HSV color information. In IEEE Intelligent Transportation Systems, pp. 334–339.
109. Zhang, Y., Zhao, C., He, J., & Chen, A. (2016). Vehicles detection in complex urban traffic scenes using Gaussian mixture model with confidence measurement. In IET ITS.
110. Lan, J., Jiang, Y., & Yu, D. (2015). A new automatic obstacle detection method based on selective updating of Gaussian mixture model. In ICTIS 2015, pp. 21–25.

111. Stauffer, C., & Grimson, E. (1999). Adaptive background mixture models for real-time tracking. In IEEE CVPR 1999, pp. 246–252.
112. Wren, C., & Azarbayejani, A. (1997, July). Pfinder: Real-time tracking of the human body. *IEEET-PAMI, 19*(7), 780–785.
113. Rodriguez, P., & Wohlberg, B. (2016). Incremental principal component pursuit for video background modeling. *Mathematical Imaging and Vision, 55*(1), 1–18.
114. Li, L., & Huang, W. (2004, November). Statistical modeling of complex background for foreground object detection. *IEEE T-IP, 13*(11), 1459–1472.
115. Zivkovic, Z., & Heijden, F. (2004). Recursive unsupervised learning of finite mixture models. *IEEE T-PAMI, 5*(26), 651–656.
116. Rodriguez, P. (2015, March). Real-time incremental principal component pursuit for video background modeling on the TK1. In GPU Technical Conference, GTC 2015.
117. Prati, A., Mikic, I., Trivedi, M., & Cucchiara, R. (2003, July). Detecting moving shadows: Algorithms and evaluation. *IEEE T-PAMI, 25*(4), 918–923.
118. KaewTraKulPong, P., & Bowden, R. (2001, September). An improved adaptive background mixture model for real-time tracking with shadow detection. In AVBS 2001.
119. Yang, T., Wang, X., Yao, B., Li, J., Zhang, Y., He, Z., & Duan, W. (2016). Small moving vehicle detection in a satellite video of an urban area. In MDPI Sensors.
120. Li, X., & Wu, Y. (2014). Image object detection algorithm based on improved Gaussian mixture model. *Journal of Multimedia, 9*(1), 152–158.
121. Lu, X., Izumi, T., Takahashi, T., & Wang, L. (2014). Moving vehicle detection based on fuzzy background subtraction. In IEEE ICFS, pp. 529–532.
122. Yuan, G., Gao, Y., Xu, D., & Jiang, M. (2011). A new background subtraction method using texture and color information. In ICIC 2011, pp. 541–558.
123. Zhang, S., Qi, Z., & Zhang, D. (2009). Ship tracking using background subtraction and interframe correlation. In CISP 2009, pp. 1–4.
124. Bloisi, D., & Iocchi, L. (2012). Independent multimodal background subtraction. In International Conference on Computational Modeling of Objects Presented in Images: *Fundamentals, Methods and Applications*, pp. 39–44.
125. Heikkila, M., & Pietikainen, M. (2006). A texture-based method for modeling the background and detecting moving objects. *IEEE T-PAMI 2006, 28*(4), 657–662.
126. Barnich, O., & Droogenbroeck, M. V. (2009, April). ViBe: a powerful random technique to estimate the background in video sequences. In International Conference on Acoustics, Speech, and Signal Processing, ICASSP 2009, pp. 945–948.
127. Huang, J., Huang, X., & Metaxas, D. (2009, October). Learning with dynamic group sparsity. In International Conference on Computer Vision, ICCV 2009.
128. Oliver, N., Rosario, B., & Pentland, A. (1999, January). A Bayesian computer vision system for modeling human interactions. In ICVS 1999.
129. McFarlane, N., & Schofield, C. (1995). Segmentation and tracking of piglets in images. In British Machine Vision and Applications, BMVA 1995, pp. 187–193.
130. Jodoin, P., Saligrama, V., & Konrad, J. (2011). Behavior subtraction. In IEEE T-IP.
131. Eng, H., Toh, K., & Yau, W. (2003). An automatic drowning detection surveillance system for challenging outdoor pool environments. In IEEE ICCV 2003, pp. 532–539.
132. Eng, H., Toh, K., Kam, A., Wang, J., & Yau, W. (2004). Novel region-based modeling for human detection within highly dynamic aquatic environment. *IEEE CVPR 2004, 2*, 390.
133. Chan, K. (2011, June). Detection of swimmer based on joint utilization of motion and intensity information. In IAPR Conference on Machine Vision Applications.
134. Chan, K. (2013, January). Detection of swimmer using dense optical flow motion map and intensity information. *Machine Vision and Applications, 24*(1), 75–101.
135. Fei, L., Xueli, W., & Dongsheng, C. (2009). Drowning detection based on background subtraction. In ICESS 2009, pp. 341–343.
136. Lei, F., & Zhao, X. (2010, October). Adaptive background estimation of underwater using Kalman-filtering. In CISP 2010.

137. Zhang, C., Li, X., & Lei, F. (2015, June). A novel camera-based drowning detection algorithm. In Chinese Conference on Image and Graphics Technologies, pp. 224–233.

138. Sha, L., Lucey, P., Sridharan, S., & Pease, D. (2014, March). Understanding and analyzing a large collection of archived swimming videos. In IEEE WACV 2014, pp. 674–681.

139. Peixoto, N., Cardoso, N., & Mendes, J. (2012). Motion segmentation object detection in complex aquatic scenes and its surroundings. In INDIN 2012, pp. 162–166.

140. Friedman, N., & Russell, S. (1997). Image segmentation in video sequences: A probabilistic approach. In UAI 1997, pp. 175–181.

141. Ridder, C., Munkelt, O., & Kirchner, H. (1995). Adaptive background estimation and foreground detection using Kalman-filtering. In ICRAM 1995, pp. 193–195.

142. Sobral, A., Bouwmans, T., & Zahzah, E. (2015). Double-constrained RPCA based on saliency maps for foreground detection in automated maritime surveillance. In ISBC 2015 Workshop conjunction with AVSS 2015, Karlsruhe, Germany, pp. 1–6.

143. Javed, S., Mahmood, A., Bouwmans, T., & Jung, S. (2017, September). Superpixels based manifold structured sparse RPCA for moving object detection. In International Workshop on Activity Monitoring by Multiple Distributed Sensing, BMVC 2017, London, UK.

144. Javed, S., Mahmood, A., Bouwmans, T., & Jung, S. (2017, December). Background-foreground modeling based on spatio-temporal sparse subspace clustering. *IEEE Transactions on Image Processing, 26*(12), pp. 5840–5854.

145. Javed, S., Mahmood, A., Bouwmans, T., & Jung, S. (2016, December). Spatiotemporal low-rank modeling for complex scene background initialization. In IEEE T-CSVT.

146. Bouwmans, T., Javed, S., Zhang, H., Lin, Z., & Otazo, R. (2018). On the applications of robust PCA in image and video processing. In Proceedings of the IEEE, pp. 1427–1457.

147. Braham, M., & Van Droogenbroeck, M. (2016, May). Deep background subtraction with scene-specific convolutional neural networks. In IWSSIP 2016, Bratislava, Slovakia.

148. Minematsu, T., Shimada, A., Uchiyama, H., & Taniguchi, R. (2018). Analytics of deep neural network-based background subtraction. In *MDPI Journal of Imaging.*

149. Liang, X., Liao, S., Wang, X., Liu, W., Chen, Y., & Li, S. (2018 July). Deep background subtraction with guided learning. In IEEE ICME 2018, San Diego, CA.

150. Sultana, M., Mahmood, A., Javed, S., & Jung, S. (2018 May). Unsupervised deep context prediction for background estimation and foreground segmentation. *Machine Vision and Applications.*

151. Laugraud, B., Pierard, S., & Van Droogenbroeck, M. (2018). LaBGen-P-Semantic: A first step for leveraging semantic segmentation in background generation. *MDPI Journal of Imaging, 4*(7), Art. 86.

152. Laugraud, B., Pierard, S., & Van Droogenbroeck, M. (2017). LaBGen: A method based on motion detection for generating the background of a scene. *Pattern Recognition Letters, 96,* 12–21.

153. Laugraud, B., Pierard, S., & Van Droogenbroeck, M. (2016). LaBGen-P: A pixel-level stationary background generation method based on LaBGen. In ICPR 2016.

154. Javed, S., Bouwmans, T., & Jung, S. (2017). SBMI-LTD: Stationary background model initialization based on low-rank tensor decomposition. In ACM SAC 2017.

155. Sobral, A., Bouwmans, T., & Zahzah, E. (2015, September). Comparison of matrix completion algorithms for background initialization in videos. In SBMI 2015, ICIAP 2015, Genoa, Italy.

156. Zhang, Y., Li, X., Zhang, Z., Wu, F., & Zhao, L. (2015, June). Deep learning driven blockwise moving object detection with binary scene modeling. *Neurocomputing, 168,* 454–463.

157. Shafiee, M., Siva, P., Fieguth, P., & Wong, A. (2016, June). Embedded motion detection via neural response mixture background modeling. In IEEE CVPR 2016.

158. Shafiee, M., Siva, P., Fieguth, P., & Wong, A. (2017, June). Real-time embedded motion detection via neural response mixture modeling. *Journal of Signal Processing Systems, 90*(6), 931–946.

159. Nguyen, T., Pham, C., Ha, S., & Jeon, J. (2018, January). Change detection by training a triplet network for motion feature extraction. In IEEE T-CSVT.

160. Varona, J., Gonzalez, J., Rius, I., & Villanueva, J. (2008). Importance of detection for video surveillance applications. In Optical Engineering, pp. 1–9.

161. Sheikh, Y., Javed, O., & Kanade, T. (2009, October). Background subtraction for freely moving cameras. In IEEE ICCV 2009, pp. 1219–1225.
162. Elqursh, A., & Elgammal, A. (2012). Online moving camera background subtraction, In European Conference on Computer Vision, ECCV 2012.
163. Sugaya, Y., & Kanatani, K. (2004). Extracting moving objects from a moving camera video sequence. In SSII 2004, pp. 279–284.
164. Taneja, A., Ballan, L., & Pollefeys, M. (2010). Modeling dynamic scenes recorded with freely. In Asian Conference on Computer Vision, ACCV 2010.

5

Visual Surveillance of Natural Environments: Background Subtraction Challenges and Methods

Thierry Bouwmans[1] and Belmar García-García[2]
[1]*Laboratoire MIA, University of La Rochelle, France*
[2]*Instituto Politécnico Nacional, Mexico City, Mexico*

CONTENTS

5.1 Introduction

For the observation of animals and insects in their natural environment, a noninvasive, simple intelligent system is necessary. Therefore, a vision-based system is a viable option to obtain information regarding animal and insect behavior in the environment. A vision-based system can be used to study social behaviors within the same group; for example, studies of bees show very close interaction with partners [1]. The system can also analyze the climate impact and changes in the ecosystem when animals, such as birds, interact with their environment [2–5]. A specific application called Fish4Knowledge (F4K) can obtain information on animal species in different climatic conditions, specifically fish behaviors to adapt to the environment in the presence of storms, sea currents, or typhoons [6–9]. According to Iwatani et al., vision-based systems can also aid in the design of robots that mimic the locomotion of certain animal species of interest [10].

Researchers in environmental development, biology, or ethology are interested in the behavior of animals that are important to maintain environmental balance, such as bees that are pollinators of flowers on the planet. Most observations of insects and animals are made in a totally natural environment; therefore, it is also of interest to detect foreign objects that do not belong to the environment and the damages that they may cause.

In this chapter we will try to analyze the main works and publications in visual surveillance in natural environments that make use of the background subtraction approach with corresponding videos and experiments whose purpose was the detection of foreground objects and background modeling for specific applications. Thus, this chapter refers only to background subtraction methods that are previously and currently used in these applications. For more details on the existing methods under the background subtraction approach, the reader can refer to many surveys in the field [11–18].

5.2 Intelligent Surveillance of Insects and Animals

Video surveillance systems for animals and insects also have their application in protected environments (forest, river, zoo, etc.) for animals and insects such as fish [6, 7], honeybees [1, 19–21], birds [2–5, 22], or hinds [23, 24]. Due to the particular characteristics that animals and insects can present in their environment in terms of behavior and visual appearance, the videos will be classified and analyzed in the following sections. Table 5.1 shows a summary of the different works published in the field, the techniques used by the authors, and characteristics of the background model and its updating,

TABLE 5.1

Visual Surveillance of Animals in Natural Environments: An Overview

Categories	Subcategories— Aims	Objects of Interest—Scenes	Authors—Dates
Intelligent visual observation of animal and insect behaviors	**Bird Surveillance**	**Birds**	
	Feeder stations in natural habitats	Feeder station webcam/ camcorder datasets	Ko et al. (2008) [2]
	Feeder stations in natural habitats	Feeder station webcam/ camcorder datasets	Ko et al. (2010) [3]
	Seabirds	Cliff face nesting sites	Dickinson et al. (2008) [4]
	Seabirds	Cliff face nesting sites	Dickinson et al. (2010) [5]
	Observation in the air	Lakes in northern Alberta (Canada)	Shakeri and Zhang (2012) [22]
	Wildlife@Home	Natural nesting stations	Goehner et al. (2015) [25]

(Continued)

TABLE 5.1 *(Continued)*

Visual Surveillance of Animals in Natural Environments: An Overview

Categories	Subcategories— Aims	Objects of Interest—Scenes	Authors—Dates
Intelligent visual observation of animal and insect behaviors *(continued)*	**Insect Surveillance**	**1) Honeybees**	
	Hygienic bees	Institute of Apiculture in Hohen-Neuendorf (Germany)	Knauer et al. (2005) [1]
	Honeybee colonies	Hive entrance	Campbell et al. (2008) [19]
	Honeybee behaviors	Flat surface—Karl-Franzens-University in Graz (Austria)	Kimura et al. (2012) [20]
	Pollen-bearing honeybees	Hive entrance	Babic et al. (2016) [21]
	Honeybee detection	Hive entrance	Pilipovic et al. (2016) [26]
		2) Spiders	
	Spider detection	Observation box	Iwatani et al. (2016) [10]
	Fish Surveillance	**Fish**	
	EcoGrid project (Taiwan)	Ken-Ding subtropical coral reef	Spampinato et al. (2008) [6]
	EcoGrid project (Taiwan)	Ken-Ding subtropical coral reef	Spampinato et al. (2010) [7]
	Fish4Knowledge (European)	Taiwan's coral reefs	Kavasidis and Palazzo (2012) [27]
	Fish4Knowledge (European)	Taiwan's coral reefs	Spampinato et al. (2014) [8, 9]
	Underwater change detection (European)	Underwater scenes (Germany)	Radolko et al. (2016) [28]
		Simulated underwater environment	Liu et al. (2016) [29]
	Rail-based fish catching	Open-sea environment	Huang et al. (2016) [30]
	Fish4Knowledge (European)	Taiwan's coral reefs	Seese et al. (2016) [31]
	Fine-grained fish recognition	Cam-trawl dataset/ chute multispectral dataset	Wang et al. (2016) [32]
	Fish4Knowledge (European)	Taiwan's coral reefs	Rout et al. (2017) [33]
Intelligent visual observation of natural environments	**River Surveillance**	**Woods**	
	Floating bottle detection	Dynamic texture videos	Zhong et al. (2003) [34]
	Floating wood detection	River videos	Ali et al. (2012) [35]
	Floating wood detection	River videos	Ali et al. (2013) [36]

color space, and foreground detection. These publications will be analyzed throughout the chapter with detailed information about animal and insect detection as well as the challenges faced in the corresponding environments.

5.2.1 Visual Observation of Birds

The importance in the constant monitoring of birds lies in the migratory phenomenon of the different species, their protection, and security in aviation. Observations can be classified as follows: (1) observations at human-made feeder stations [2, 3], (2) observation at natural nesting stations [4, 5, 25], and (3) aerial observations focusing the camera towards lakes or the roofs of buildings [22]. For the first scenario, birds at feeder stations, Ko et al. [2] found that background pixels show a high variation in their values due to the movement and presence of these animals. On the other hand, rapid background updating was not appropriate in situations where the birds presented slow movements and often were incorporated into the background.

In the case of bird nesting, Goehner et al. [25] carried out important studies for the Wildlife@Home project in the detection of important events in outdoor environments. The challenges to be addressed were a robust background model able to quickly compensate for the lighting problems of the camera and in turn the detection of small- and medium-sized cryptic birds. To solve these problems, Goehner et al. [25] used the background models mixture of Gaussian (MoG) [37], ViBe [38], and pixel-based adaptive segmentation (PBAS) [39]. ViBe and PBAS were modified to work with the second frame and morphological opening and closing filters were added to handle the noise present in the videos. Additionally, a convex hull was added around the connected foreground region to compensate for the cryptic colors of the birds.

5.2.2 Visual Observation of Fish

The analysis for fish video surveillance has its application in the following contexts: (1) detection and recognition of species in the open sea by Wang et al. [32] and in indoor environments such as tanks by El Baf et al. [40, 41]; (2) for aquaculture applications, detection and tracking for counting in a tank as in Abe et al. [42]; (3) the study of species in different weather conditions, in the open sea as in Spampinato et al. [6–9]; and (4) estimation of size and catch of fish in fishing applications as in Huang et al. [30]. All these applications face challenges because fish are deformable and flexible objects, that is, not rigid, with different appearances and colors; thus, fish identification as foreground objects is not a trivial task. Scientific literature shows that in 2007 for the Aqu@theque project, Baf et al. [40, 41] studied the kernel density estimation (KDE) [43], Mo) [44], and SG [45] background models and concluded that KDE is slower than MoG and SG for real-time applications due to its high consumption of memory resources, making MoG the most suitable for the detection of fish. For the EcoGrid project, in 2008, Spampinato et al. [6, 8] made use of the Zivkovic-Heijden Gaussian mixture model (GMM) [46] and the moving average for fish detection. In the year 2014, Spampinato et al. [7] created a texton-based KDE model to deal with erratic movement problems that fish generate randomly in different directions. Liu et al. [29], for autonomous underwater vehicles (AUVs) or remote-operated vehicles (ROVs), combined the segmentation results by the running average and the three computational fluid dynamics (CFD) with a logical "AND." Huang et al. [30] combined trajectories in movement with background subtraction to face marine surface-noise problems. The fish are detected and separated from the noise based on their trajectories from the front masks obtained by SuBSENSE [47].

5.2.3 Visual Observation of Dolphins

Ethologists and ecologists are interested in analyzing the behavior of marine mammals such as dolphins inside and outside their natural environments. The continuous monitoring of this species helps researchers to better understand their social dynamics as well as their interaction with humans. Technologies such as multiple video cameras are a resource that allows researchers to carry out long-term tasks for the detection and tracking of dolphins. Works such as Karnowski et al. [48] obtained statistical information related to the paths that the dolphins used for a long period using the background subtraction approach.

5.2.4 Visual Observation of Honeybees

The filming of honeybees is usually done at the entrance of the hive in order to count and track different objectives as explained in the following points: (1) detection of external agents to the honeybee as in Knauer et al. [1], (2) continuous monitoring at the entrance of the beehive (arrivals and departures) as in Campbell et al. [19], (3) the interaction among members of the hive as in Kimura et al. [20], and (4) monitoring of pollination as in Babic et al. [21]. As explained in Campbell et al. [19] and in Babic et al. [21], there are particular problems in the area of computer vision for the detection of bees. The portion to be detected for this target within each frame is very small, with almost identical appearance, with the number of pixels at approximately 6×14 pixels. High-resolution cameras or multiple cameras can be used to improve detection, but the computational cost or infrastructure increases. Regarding the dynamics, it is difficult to follow an object between frames due to the honeybees' fast and chaotic movements.

Honeybees change drastically between modes of flight or movement (flying, loitering, and crawling); this fact complicates the use of a unimodal movement model. Weather conditions, lighting conditions, and the season of the year are factors that complicate the detection process. Other factors to take into account include shadows due to the chamber enclosure and the movement of the trees caused by the wind. In addition, hive placement is restricted in relation to buildings, the environment, or compass points. The installation of artificial lighting could affect the natural behavior of honeybees.

The scenes at the entrance of the hives are usually agglomerated, with many occlusions, groupings, or overlaps, which causes groups of honeybees to be detected as a single object of interest. When it is not possible to segment individually, it is enough to know if any group of honeybees is a carrier of pollen [21]. In 2016, Pilipovic et al. [26] concluded that the MoG is the most suitable algorithm for the detection of honeybees at the hive entrances, after analyzing the Kalman filter [49], frame differencing, median model [50], and MoG models [44] applied to this field.

5.2.5 Visual Observation of Spiders

A hunting robot design was proposed by Iwatani et al. [10] to mimic the behavior of wolf spiders. The experiments were carried out in a controlled environment, making use of a simple background subtraction method to estimate the position and direction of the insect. One pixel per image of the grayscale video sequence was taken as the candidate component of the spider to determine if the difference between the background image and the recent frames exceeded a certain threshold.

5.2.6 Visual Observation of Lizards

Nguwi et al. [51] carried out lizard detection using a simple segmentation method with different thresholds. They discussed the use of image processing technique to detect lizards, and defragmented the video into a total of 3459 images; some are with only background scene, some contain lizards. They applied background subtraction to segment out the lizard, followed by experiments comparing thresholding values and methods, achieving an encouraging average hit rate of 98% and average computation time of 1.066 seconds.

5.2.7 Visual Observation of Pigs

Analyzing the behavior of pigs on a farm is a task that allows farmers to provide better living conditions to animals by controlling in an appropriate way the environment's temperature and the pigs' interactions with humans. Basically the three challenges to be solved in the farrowing pens according to Tu et al. [52] are:

- **Multimodal background:** The materials existing in the farrowing pen are detected as a dynamic background such as the movement of the pig or the piglets.
- **Illumination changes:** The most serious problem occurs when the lights in the farrowing pen turn off or on, causing the entire frame to show foreground detection. This situation is common in most statistical models when there is a sudden change of illumination in the whole scene.
- **Static foregrounds:** The pigs are incorporated into the background model during long periods of sleep.

Guo et al. [53] detected pigs from aerial views using a prediction mechanism-mixture of Gaussians (PM-MoG) model. Tu et al. [52] combined the dyadic wavelet transform (DWT) and a modified MoG model. This algorithm proved to be robust in complex environments by accurately segmenting the silhouette of a sow. Tu et al. [54] used a homomorphic wavelet filter (HWF) and a wavelet quotient image (WQI) model based on DWT for the estimation of illumination and reflectance. Based on this, Tu et al. [54] used the CFD algorithm of Li and Leung [55] to detect sows in grayscale video combining textures and intensity differences.

5.2.8 Visual Observation of Hinds (Female Deer)

Basically there are three challenges to be solved to detect hinds in the forests according to Khorrami et al. [23]:

- **Camouflage:** Hinds often use the forest background to confuse predators.
- **Static foregrounds:** The hinds are incorporated into the background model during long periods of sleep.
- **Fast movement:** Hinds move at high speeds to flee from predators.

Giraldo-Zuluaga et al. [24, 56] carried out detection of hinds in the forests of Colombia using a multilayer robust principal component analysis (RPCA). Experimental tests [24] against other RPCA models show the robustness of the multilayer RPCA model when facing complex challenges such as illumination changes.

Tables 5.2 summarizes the different works published in the field, the techniques used by the authors, and the characteristics of the background model and its updating, color space,

TABLE 5.2

Visual Surveillance of Animals: Background Models

Animal and Insect Behaviors	Type	Background Model	Background Maintenance	Foreground Detection	Color Space	Strategies
Fish surveillance	1) Tank environment					
	1.1) Species identification					
	Penciuc et al. (2006) [57]	MoG [44]	Idem MoG	Idem MoG	RGB	—
	Baf et al. (2007) [40, 41]	MoG [44]	Idem MoG	Idem MoG	RGB	—
	1.2) Industrial Aquaculture					
	Pinkiewicz (2012) [58]	Average/median	Idem average/ Idem median	Idem average/ Idem median	RGB	—
	Abe et al. (2017) [53]	Average	Idem average	Idem average	RGB	—
	Zhou et al. (2017) [59]	Average	Idem average	Idem average	Near infrared	—
	2) Open-sea environment					
	Spampinato et al. (2008) [6]	Sliding average/ Zivkovic-Heijden GMM [46]	—	—	—	AND
	Spampinato et al. (2010) [8]	Sliding average/ Zivkovic-Heijden GMM [46]	—	—	—	AND
	Spampinato et al. (2014) [9]	GMM, [44] APMM, [60] IM, [61] Wave-Back [62]	—	—	—	Fish detector
	Spampinato et al. (2014) [7]	Texton-based KDE [7]				
	Liu et al. (2016) [29]	Running average-three CFD [29]	Idem RA-TCFD	Idem RA-TCFD	RGB	AND
	Huang et al. (2016) [30]	SuBSENSE [47]	Idem SuBSENSE	Idem SuBSENSE	RGB-LBSP [53]	Trajectory feedback
	Seese et al. (2006) [31]	MoG [54]/ Kalman filter [64]	Idem MoG/ Kalman filter	Idem MoG/ Kalman filter	Intensity	Intersection
	Wang et al. (2006) [32]	MoG [44]	Idem GMM	Idem GMM	RGB	Double local thresholding
Dolphin surveillance	Karnowski et al. (2015) [48]	RPCA [65]	—	—	Intensity	—

(Continued)

TABLE 5.2 *(Continued)*

Visual Surveillance of Animals: Background Models

Animal and Insect Behaviors	Type	Background Model	Background Maintenance	Foreground Detection	Color Space	Strategies
Honeybee surveillance	Knauer et al. (2005) [1]	K-Clusters [66]	Idem K-Clusters	Idem K-Clusters	Intensity	—
	Campbell et al. (2008) [19]	MoG [44]	Idem MoG	Idem MoG	RGB	—
	Kimura et al. (2012) [20]	—	—	—	—	—
	Babic et al. (2016) [21]	MoG [44]	Idem MoG	Idem MoG	RGB	—
	Pilipovic et al. (2016) [26]	MoG [44]	Idem MoG	Idem MoG	RGB	—
Spider surveillance	Iwatani et al. (2016) [10]	Image without foreground objects	—	Threshold	Intensity	—
Lizard surveillance	Nguwi et al. (2016) [51]	—	—	—	—	—
Pig surveillance	Guo et al. (2015) [53]	PM-MoG [53]	Idem MoG	Idem MoG	RGB	Prediction mechanism
	Tu et al. (2014) [52]	MoG-DWT [52]	Idem MoG	Idem MoG	Intensity/ texture	OR
	Tu et al. (2015) [54]	CFD [55]	—	Difference	RGB	Illumination and reactance Estimation
Hind surveillance	Khorrami et al. (2012) [23]	RPCA [65]	—	—	Intensity	—
	Giraldo-Zuluaga et al. (2017) [24]	Multilayer RPCA [24]	—	—	RGB	—
	Giraldo-Zuluaga et al. (2017) [56]	Multilayer RPCA [24]	—	—	RGB	—
Bird surveillance	**Birds**					
	1) Feeder stations	KDE [67]	Blind maintenance	Bhattacharyya distance	RGB	Temporal consistency
	Ko et al. (2008) [2]	Set of warping layer [3]	Blind maintenance	Bhattacharyya distance	UYV	—
	Ko et al. (2010) [3]					
	2) Birds in air	Zivkovic-Heijden GMM [46]	Idem GMM	Idem GMM	RGB	Correspondence component
	Shakeri and Zhang (2012) [22]					
						Based on point-tracking

(Continued)

TABLE 5.2 *(Continued)*

Visual Surveillance of Animals: Background Models

Animal and Insect Behaviors	Type	Background Model	Background Maintenance	Foreground Detection	Color Space	Strategies
Bird surveillance *(continued)*	Nazir et al. (2017) [68]	OpenCV background subtraction (WiseEye)	—	—	—	—
3) Avian nesting	MoG [37], ViBe [38], PBAS [39]	Idem MoG/ PBAS/ViBe	Idem MoG/ PBAS/ViBe	RGB	Morphological processing
	Goehner et al. (2015) [25]	MoG [4]	Idem MoG	Idem MoG	RGB	Spatially coherent segmentation [69]
	Dickinson et al. (2008) [4]	MoG [4]	Idem MoG	Idem MoG	RGB	Spatially coherent segmentation [69]
	Dickinson et al. (2010) [5]					

and foreground detection. Therefore, there is no general statistical model of background subtraction for different animals and environments due to the variety of intrinsic characteristics such as shapes, sizes, textures, and behaviors. The events, International Conference on Pattern Recognition (ICPR), and visual observation and analysis of animal and insect behavior (VAIB) addressed the challenges in the detection of animals and insects.

5.3 Intelligent Visual Surveillance in Natural Environments

In this field, MotionMeerkat [70] is a general-purpose tool for motion detection in ecological environments. Its operation is based on MoG or running Gaussian average (RGA); therefore, the camouflage and abrupt illumination changes are still problems to solve. Another tool called DeepMeerkat was proposed by Weinstein [71] for background modeling based on convolutional neural network (CNN). There is a computer tool for marine environments called video and image analytics for a marine environment (VIAME). Unfortunately, this framework lacks the background subtraction algorithms necessary for complex environments. In real settings such as rivers, forests, or oceans, scene backgrounds are dynamic and difficult to predict. Sometimes objects of interest tend to camouflage themselves within the environment.

5.3.1 Forest Environments Surveillance

In forests, the foreground objects to be detected correspond to animals or humans. The challenges to be solved are the partial occlusions of the objects of interest and the abrupt transitions between the light and shadows caused by the continuous movement of the foliage of the trees. Boult et al. [72] conducted experiments using an omnidirectional camera to detect humans. Subsequently, Shakeri and Zhang [73] used a camera trap to detect animals in clearing zones of the forest implementing a robust PCA method proving to be better than most previous PCAs in the illumination change dataset (ICD) [73].

Yousif et al. [74] proposed a deep learning classifier in conjunction with a background model to detect people and animals. To be robust to the agitation of trees caused by the wind, the model is based on four characteristics: local binary pattern, histogram, histogram of oriented gradient (HOG), and intensity. Janzen et al. [75] carried out foreground segmentation using image thresholding, fragment histogram analysis, and background subtraction. The experiments demonstrated a human detection success rate between 60% and 92% with a reduction of one-third the number of input frames.

5.3.2 River Environments Surveillance

The main objective is to detect strange elements in the river such as glass bottles, floating pieces of wood, and garbage. Reasons for this include protecting the river as a natural environment, and preserving structures such as dams and bridges [35, 36]. In the first situation, the idea is to keep the environment in good condition and protect the fauna [34]. In the second situation, pieces of wood, ranging from small branches or shrubs to fallen trees, can damage the dams and bridges. The damage caused is proportional to the size of the element considered foreign. Fallen trees on the pillars of bridges represent a greater danger to the river, as these will cause accumulation of debris, shrubs, and branches. The river environment can be challenging because both the water and the objects of interest are in motion. In flood situations, the water flow varies considerably causing more movement. Additional variables in the background model such as cloud shadows and water-surface lighting should be considered. With the goal of protecting the river as a natural environment, Zhong et al. [34] applied the Kalman filter in a waving river to detect bottles (2003). In order to find ways to protect structures such as dams and bridges, Ali et al. [35] applied a space-time spectral mode (2012). In another publication, to detect floating wood, Ali et al. [36] added to the MoG model a rigid motion model.

5.3.3 Ocean Environments Surveillance

The main objectives in surveillance of ocean environments are: (1) in most cases the detection of ships for the optimization of maritime traffic, (2) timely detection of foreign objects to avoid a collision [76], and (3) prevention of pirate attacks on ships in both port areas or the open sea by detecting intruders (unknown persons) [77]. The most challenging scenes are on the coasts due to the breaking of waves. Normally the movement of water leads to false-positive detections in segmentation methods [78]. Climate changes and boat wakes contribute to the generation of highly dynamic scene backgrounds.

Prasad et al. [79, 80] contributed a list of challenges from the evaluation of the 34 ChangeDetection.net algorithms applied to the Singapore-Maritime dataset. The experiments [79] demonstrated the ineffectiveness of all these methods by producing false-positive and false-negative rates when considering water as a foreground object or objects of interest as part of the background. Conditions of a real environment in this type of scenario such as darkness, fog, rain, sunlight reflections, weather conditions, brightness in the water, and the tides contribute to the dynamism of an oceanic environment.

The statistical distribution of water and sea, contrast, and visibility of objects at a great distance can be affected by natural conditions of the environment. For example, speckle and glint effects cause the background statistical distribution to be nonuniform and the objects of interest to be incorporated into the background, or vice versa. The range of colors for light conditions vary significantly: hazy conditions (predominantly gray), bright daylight (predominantly blue), sunset (predominantly yellow and red), and night (predominantly dark).

5.3.4 Submarine Environments Surveillance

Underwater environments can be classified in three ways: (1) swimming pools, (2) tank environments for fish, and (3) open-sea environments.

5.3.4.1 Swimming Pools

Challenges to be faced in swimming pools include specular reflections, splashes, water ripples, and rapid changes in lighting. As a first work, Eng et al. [81] proposed modeling the background of this scenario using block-based median in the CIELAB color space to detect swimmers in outdoor environments. A Markov random field framework improved the tracking of the objects of interest due to partial occlusions. In another work, Eng et al. [82] used the CIELAB color space for the detection of swimmers, similar to the work by Eng et al. [81]. The background is characterized by a region-based single-multivariate Gaussian model. Additionally, a spatio-temporal filtering compensated the occlusion problems caused by specular reflections.

On the other hand, to face the problems of water ripple and light spot, Lei and Zhao [83] used a Kalman filter [64]. Chan et al. [84] applied the optical flow algorithm and MoG to detect swimmers and different swimming styles (backstroke, freestyle, and breaststroke) in daytime and nighttime videos. Peixoto et al. [85] used the HSV color space and worked with the CFD combined with the average model to estimate the background of the environment.

5.3.4.2 Tank Environments

Tanks can be classified in three ways: (1) tanks for industrial purposes such as aquaculture, (2) tanks for the study of marine species, and (3) tanks that reproduce the maritime natural environment such as aquariums. The challenges to face in this type of environment can be classified as:

- **Environmental challenges:** These include movement due to the continuous renewal of the water and the movement caused by the fish. Another challenge is illumination changes in the natural environment, and those due to artificial lighting inside and outside the tank. Moreover, the movement of algae in aquariums causes detection problems as demonstrated in Baf et al. [40, 41].

- **Fish intrinsic features:** The movement of fish is affected by the type of species and the shape of the tank. The number of fish inside the tank is another factor to be taken into account for detection process. The behavior of fish of the same species is similar even when several species interact within the same tank, according to Baf et al. [40]. Fish species swim at different marine depths. Frequently, the targets can be partially obstructed by algae or other fish species that are not of interest. In aquaculture applications, the number of fish in a tank is larger and of the same species than in an aquarium. Therefore, there is a more or less similar behavior. Abe et al. [42] achieved the detection of 170–190 fish of 250 Japanese rice fish because this species tends to swim at different depths.

5.3.4.3 Open-Sea Environments

Water can have different degrees of clarity or cleanliness. Luminosity depends on factors such as time of day and weather conditions as described in Spampinato et al. [6]. In subtropical waters, the growth of algae in the lenses of the cameras generates videos with greenish or bluish tones. Therefore, cleaning and constant maintenance of the recording equipment is necessary. According to Kavasidis and Palazzo [27] and Spampinato et al. [9], the problems and challenges present in open-sea environments are classified as:

- **Illumination changes:** Because video recordings are made throughout the day, background subtraction algorithms must take into account all possible conditions and light transitions.

- **Physical phenomena:** The quality of the image acquired in terms of contrast and video clarity is compromised by natural phenomena such as typhoons, sea currents, and storms.

- **Murky water:** The detection and monitoring of fish can be affected by the clarity of the water due to the presence of plankton or drift. This causes the detection of false positives to objects that do not correspond to being fish.

- **Grades of freedom:** In a vehicular- or pedestrian-traffic video, moving objects are confined to two dimensions. However, fish in underwater videos can move in three dimensions.

- **Presence of algae in camera lens:** Formation of algae in the lens of the camera as a result of contact with sea water is a challenge.

- **Periodic and dynamic background:** Objects with arbitrary and periodic movement that form part of the background, such as stones and plants subject to tide and drift, cause the detection of false positives in the segmentation methods.

In 2012, Kavasidis and Palazzo [27] evaluated on the Fish4Knowledge dataset [86] the performance of six background subtraction algorithms in the task of fish detection in unconstrained and underwater video (ViBe [38], GMM [44], IM [61], APMM [60], Wave-Back [62], and Codebook [87]). The comparative evaluation showed: (1) At the blob level, under normal weather conditions, static background, and constant illumination, an acceptable performance was observed except for the Wave-Back algorithm. GMM and APMM occupied intermediate positions, with GMM being slightly higher than APMM. The ViBe algorithm proved to be the best in most videos, whereas the Codebook algorithm showed better results in high-resolution videos. (2) At the pixel level, all algorithms proved to be able to identify in an acceptable manner the pixels corresponding to the foreground, for example, a detection rate of 83.2% for the APMM algorithm and 83.4% for the ViBe algorithm. The false-alarm rate is high in most cases, especially in the IM and Wave-Back algorithms when the video contrast is low, when scenes are in low light, or in weather conditions such as storms or typhoons.

In another work in 2017, Radolko et al. [28] found five challenges: (1) Blurring due to forward scattering in the water makes it impossible to obtain a clear image. (2) Small particles in the water cause back scatter. The effect is similar to a veil in front of the scene. (3) Water absorbs light more strongly than air. In addition, the absorption effect depends on the wavelength of the light, and this leads to underwater images with very distorted and mitigated colors. (4) Ripples on the surface of the water cause light reflections in the ground. They are similar to strong shadows, and fast movements make them very difficult to differentiate from dark objects. (5) Small floating particles strongly reflect light. For the most part, they are small enough to filter during the segmentation process; however, they still damage the image and complicate, for example, the static background modeling.

In experimental tests carried out on UnderwaterChangeDetection.eu [28], GSM [88] demonstrated a better performance than EA-KDE (also known as KNN) [89], MoG-EM [90], and Zivkovic-Heijden GMM (also known as ZHGMM) [46]. In another work, the spatio-contextual GMM (SC-GMM) designed by Rout et al. [33] proved to be better than 18 background subtraction algorithms; classical algorithms such as the original KDE, MoG, and PCA; and advanced algorithms such as PBAS [39], DPGMM (VarDMM) [91], PBAS-PID [92], SOBS [93], and SuBSENSE [47]. The tests were performed in the Fish4Knowledge dataset [86] and the dataset UnderwaterChangeDetection.eu [28]. From this analysis, uncontrolled external environments imply the existence of multimodal backgrounds; therefore, a robust background model must be able to adapt to the changes suffered by the scene over time to reduce the false-detection rate.

The problems of detection in oceanic surfaces and in submarine environments were addressed by Computer Vision for Analysis of Underwater Imagery (CVAUI) in 2015 and later by ICPR in 2016. Each application presents intrinsic characteristics in terms of objects of interest and the environment. These elements work as a base to determine the technique of background subtraction tailored to a specific need. The availability of more or less recent publications represents another additional factor to address the problem.

5.4 Solved and Unsolved Challenges and Prospective Solutions

Table 5.3 groups the solved and unsolved challenges, ordered by applications, in order to provide information on the needs for future research. The main unsolved challenges are multimodal backgrounds, illumination changes, and the low frame rate for camera traps. To handle these challenges, recent background subtraction methods based on RPCA [94–98] and deep learning [99–102] can be employed. In addition, robust background initialization methods [103–107] as well as robust deep learning features [108–111] should be considered for very complex environments such as maritime environments.

TABLE 5.3

Solved and Unsolved Challenges in Visual Surveillance of Human Activities

Applications	Scenes	Challenges	Solved/ Unsolved
1) Intelligent Visual Observation of Animal and Insect Behaviors			
1.1) Bird surveillance	Outdoor scenes	Multimodal backgrounds	Partially solved
		Illumination changes	Partially solved
		Camera jitter	Partially solved
1.2) Fish surveillance	Aquatic scenes	Multimodal backgrounds	Partially solved
		Illumination changes	Partially solved
1.3) Dolphin surveillance	Aquatic scenes	Multimodal backgrounds	Partially solved
		Illumination changes	Partially solved
1.4) Honeybee surveillance	Outdoor scenes	Small objects	Partially solved
1.5) Spider surveillance	Outdoor scenes		Partially solved
1.6) Lizard surveillance	Outdoor scenes	Multimodal backgrounds	Partially solved
1.7) Pig surveillance	Indoor scenes	Illumination changes	Partially solved
1.8) Hind surveillance	Outdoor scenes	Multimodal backgrounds	Partially solved
		Low frame rate	Partially solved
2) Intelligent Visual Observation of Natural Environments			
2.1) Forest	Outdoor scenes	Multimodal backgrounds	Partially solved
		Illumination changes	Partially solved
		Low frame rate	Partially solved
2.2) River	Aquatic scenes	Multimodal backgrounds	Partially solved
		Illumination changes	Partially solved
2.3) Ocean	Aquatic scenes	Multimodal backgrounds	Partially solved
		Illumination changes	Partially solved
2.4) Submarine	Aquatic scenes	Multimodal backgrounds	Partially solved
		Illumination changes	Partially solved

5.5 Conclusion

In this chapter, we first surveyed the main visual surveillance applications of natural environments to detect static or moving objects of interest, where a background subtraction approach is used to model the background of the scene. Then, we reviewed the challenges related to the different environments and the different moving objects. In all these applications, other moving object detection is crucial as it is the first step that is followed by tracking, recognition, or behavior analysis. So, the foreground mask must provide high accuracy even in the presence of challenging environments such as marine environments. Different critical situations appear following the applications as (1) the location of the cameras that can detect small or large moving objects with respect to the size of the images, (2) the type of the environments, and (3) the type of the animals.

Due to the variety of environments and animals with very different intrinsic properties in terms of appearance, it is necessary to develop specific background models for specific applications or to find a universal background model that can be used in all applications. The best solution may be to develop a dedicated background model for particular challenges, and to pick up the suitable background model when the corresponding challenges are detected. For natural environments, statistical models offer a suitable framework but challenges such as dynamic backgrounds, illumination changes, and sleeping/beginning foreground objects require additional specific developments. Maritime and aquatic environments require more robust methods than the top methods of ChangeDetection.net competition [79]. In addition, background subtraction can be also used in applications in which cameras are slowly moving [112–115].

References

1. Knauer, U., Himmelsbach, M., Winkler, F., Zautke, F., Bienefeld, K., & Meffert, B. (2005). Application of an adaptive background model for monitoring honeybees. In VIIP 2005.
2. Ko, T., Soatto, S., & Estrin, D. (2008, October). Background subtraction on distributions. In European Conference on Computer Vision, ECCV 2008, pp. 222–230.
3. Ko, T., Soatto, S., & Estrin, D. (2010, June). Warping background subtraction. In IEEE CVPR 2010.
4. Dickinson, P., Freeman, R., Patrick, S., & Lawson, S. (2008). Autonomous monitoring of cliff nesting seabirds using computer vision. In International Workshop on Distributed Sensing and Collective Intelligence in Biodiversity Monitoring.
5. Dickinson, P., Qing, C., Lawson, S., & Freeman, R. (2010). Automated visual monitoring of nesting seabirds. In ICPRW, VAIB 2010.
6. Spampinato, C., Burger, Y., Nadarajan, G., & Fisher, R. (2008). Detecting, tracking and counting fish in low quality unconstrained underwater videos. In VISAPP 2008, pp. 514–519.
7. Spampinato, C., Palazzo, S., & Kavasidis, I. (2014). A texton-based kernel density estimation approach for background modeling under extreme conditions. *Computer Vision and Image Understanding, CVIU 2014, 122,* 74–83.
8. Spampinato, C., Giordano, D., & Nadarajan, G. (2010). Automatic fish classification for underwater species behavior understanding. In ACM ARTEMIS 2010, pp. 45–50.
9. Spampinato, C., Palazzo, S., Boom, B., van Ossenbruggen, J., Kavasidis, I., Salvo, R. D. ... Fisher, R. (2014). Understanding fish behavior during typhoon events in real-life underwater environments. In Multimedia Tools and Applications, pp. 1–38.
10. Iwatani, Y., Tsurui, K., & Honma, A. (2016, December). Position and direction estimation of wolf spiders, pardosa astrigera, from video images. In ICRB 2016.
11. Bouwmans, T., Baf, F. E., & Vachon, B. (2010, January). Statistical background modeling for foreground detection: A survey. In *Handbook of Pattern Recognition and Computer Vision* (pp. 181–199). World Scientific Publishing.
12. Bouwmans, T. (2012, March). Background subtraction for visual surveillance: A fuzzy approach. In *Handbook on Soft Computing for Video Surveillance* (pp. 103–139). Taylor and Francis Group.
13. Bouwmans, T., Baf, F. E., & Vachon, B. (2008, November). Background modeling using mixture of Gaussians for foreground detection—a survey. *RPCS 2008, 1*(3), 219–237.

14. Bouwmans, T. (2009, November). Subspace learning for background modeling: a survey. *Recent Patents on Computer Science, RPCS 2009, 2*(3), 223–234.
15. Bouwmans, T. (2011, September). Recent advanced statistical background modeling for foreground detection: a systematic survey. *RPCS 2011, 4*(3), 147–176.
16. Bouwmans, T., & Zahzah, E. (2014, May). Robust PCA via principal component pursuit: A review for a comparative evaluation in video surveillance. *Special Issue on Background Models Challenge, CVIU 2014, 122*, 22–34.
17. Bouwmans, T. (2014, May). Traditional and recent approaches in background modeling for foreground detection: An overview. *Computer Science Review, 11*, 31–66.
18. Bouwmans, T., Sobral, A., Javed, S., Jung, S., & Zahzah, E. (2017, February). Decomposition into low-rank plus additive matrices for background/foreground separation: A review for a comparative evaluation with a large-scale dataset. *Computer Science Review, 23*, 1–71.
19. Campbell, J., Mummert, L., & Sukthankar, R. (2008, December). Video monitoring of honey bee colonies at the hive entrance. In ICPRW, VAIB 2008.
20. Kimura, T., Ohashi, M., Crailsheim, K., Schmickl, T., Odaka, R., & Ikeno, H. (2012, November). Tracking of multiple honey bees on a flat surface. In ICETET 2012, pp. 36–39.
21. Babic, Z., Pilipovic, R., Risojevic, V., & Mirjanic, G. (2016, July). Pollen bearing honey bee detection in hive entrance video recorded by remote embedded system for pollination monitoring. In ISPRS Congress.
22. Shakeri, M., & Zhang, H. (2012, July). Real-time bird detection based on background subtraction. In World Congress on Intelligent Control and Automation, WCICA 2012, pp. 4507–4510.
23. Khorrami, P., Wang, J., & Huang, T. (2012). Multiple animal species detection using robust principal component analysis and large displacement optical flow. In ICPRW 2012, VAIB 2012.
24. Giraldo-Zuluaga, J., Gomez, A., Salazar, A., & Diaz-Pulido, A. (2019). Camera-trap images segmentation using multi-layer robust principal component analysis. *The Visual Computer, 35*(3), 335–347.
25. Goehner, K., Desell, T., et al. (2015). A comparison of background subtraction algorithms for detecting avian nesting events in uncontrolled outdoor video. In IEEE International Conference on e-Science, pp. 187–195.
26. Pilipovic, R., Risojevic, V., Babic, Z., & Mirjanic, G. (2016, March). Background subtraction for honey bee detection in hive entrance video. In INFOTEH-JAHORINA, 15.
27. Kavasidis, I., & Palazzo, S. (2012). Quantitative performance analysis of object detection algorithms on underwater video footage. In ACM MAED 2012, pp. 57–60.
28. Radolko, M., Farhadifard, F., & von Lukas, U. (2016). Dataset on underwater change detection. In IEEE Monterey OCEANS 2016.
29. Liu, H., Dai, J., Wang, R., Zheng, H., & Zheng, B. (2016). Combining background subtraction and three-frame difference to detect moving object from underwater video. In OCEANS 2016, pp. 1–5.
30. Huang, T., Hwang, J., Romain, S., & Wallace, F. (2016). Live tracking of rail-based fish catching on wild sea surface. In ICPRW 2016, CVAUI 2016.
31. Seese, N., Myers, A., Smith, K., & Smith, A. (2016). Adaptive foreground extraction for deep fish classification. In ICPRW 2016, CVAUI 2016, pp. 19–24.
32. Wang, G., Hwang, J., Williams, K., Wallace, F., & Rose, C. (2016). Shrinking encoding with two-level codebook learning for fine-grained fish recognition. In ICPRW 2016, CVAUI 2016.
33. Rout, D., Subudhi, B., Veerakumar, T., & Chaudhury, S. (2017). Spatio-contextual Gaussian mixture model for local change detection in underwater video. In ESWA.
34. Zhong, J., & Sclaroff, S. (2003). Segmenting foreground objects from a dynamic textured background via a robust Kalman filter. In IEEE ICCV 2003, pp. 44–50.
35. Ali, I., Mille, J., & Tougne, L. (October, 2012). Space-time spectral model for object detection in dynamic textured background. *IEEE AVSS 2010, 33*(13), 1710–1716.
36. Ali, I., Mille, J., & Tougne, L. (2013, July). Adding a rigid motion model to foreground detection: Application to moving object detection in rivers. In PAA.

37. Power P. & Schoonees, J. (2002, November). Understanding background mixture models for foreground segmentation. In Imaging and Vision Computing New Zealand.
38. Barnich, O., & Droogenbroeck, M. V. (2009, April). ViBe: a powerful random technique to estimate the background in video sequences. In IEEE ICASSP 2009, pp. 945–948.
39. Hofmann, M., Tiefenbacher, P., & Rigoll, G. (2012, June). Background segmentation with feedback: The pixel-based adaptive segmenter. In IEEE CDW 2012, CVPR 2012.
40. Baf, F. E., Bouwmans, T., & Vachon, B. (2007, June). Comparison of background subtraction methods for a multimedia application. In IWSSIP 2007, pp. 385–388.
41. Penciuc, D., Baf, F. E., & Bouwmans, T. (2006, September). Comparison of background subtraction methods for an interactive learning space. In NETTIES 2006.
42. Abe, S., Takagi, T., Takehara, K., Kimura, N., Hiraishi, T., Komeyama, K., ... Asaumi, S. (2016). How many fish in a tank? Constructing an automated fish counting system by using PTV analysis. In International Congress on High-Speed Imaging and Photonics.
43. Elgammal, A., Harwood, D., & Davis, L. (1999, September). Non-parametric model for background subtraction. In Frame Rate Workshop, IEEE ICCV 1999.
44. Stauffer, C., & Grimson, E. (1999). Adaptive background mixture models for real-time tracking. In IEEE CVPR 1999, pp. 246–252.
45. Wren, C., & Azarbayejani, A. (1997, July). Pfinder: Real-time tracking of the human body. *IEEE Transactions on Pattern Analysis and Machine Intelligence, 19*(7), 780–785.
46. Zivkovic, Z., & Heijden, F. (2004). Recursive unsupervised learning of finite mixture models. *IEEE Transaction on Pattern Analysis and Machine Intelligence, 5*(26), 651–656.
47. St-Charles, P., Bilodeau, G., & Bergevin, R. (2014, June). Flexible background subtraction with self-balanced local sensitivity. In IEEE Change Detection Workshop, CDW 2014.
48. Karnowski, J., Hutchins, E., & Johnson, C. (2015). Dolphin detection and tracking. In IEEE Winter Applications and Computer Vision Workshops, WACV 2015, pp. 51–56.
49. Karmann, K., & Brand, A. V. (1990). Moving object recognition using an adaptive background memory. In *Time-Varying Image Processing and Moving Object Recognition*. Elsevier.
50. McFarlane, N., & Schofield, C. (1995). Segmentation and tracking of piglets in images. In British Machine Vision and Applications, BMVA 1995, pp. 187–193.
51. Nguwi, Y., Kouzani, A., Kumar, J., & Driscoll, D. (2016). Automatic detection of lizards. In ICAMechS 2016, pp. 300–305.
52. Tu, G., Karstoft, H., Pedersen, L., & Jorg, E. (2014, January). Segmentation of sows in farrowing pens. In IET Image Processing.
53. Guo, Y., Zhu, W., Jiao, P., & Chen, J. (2014, September). Foreground detection of group-housed pigs based on the combination of mixture of Gaussians using prediction mechanism and threshold segmentation. *Biosystems Engineering, 125*, 98–104.
54. Tu, G., Karstoft, H., Pedersen, L., & Jorg, E. (2015). Illumination and reflectance estimation with its application in foreground detection. *MDPI Sensors 2015, 15*, 21407–21426.
55. Li, L., & Leung, M. (2002, February). Integrating intensity and texture differences for robust change detection. *IEEE Transactions on Image Processing, 11*, 105–112.
56. Giraldo-Zuluaga, J., Salazar, A., Gomez, A. & Diaz-Pulido, A. (2017). Automatic recognition of mammal genera on camera-trap images using multi-layer robust principal component analysis and mixture neural networks.
57. Baf, F. E., & Bouwmans, T. (2007, July). Comparison of background subtraction methods for a multimedia learning space. In SIGMAP 2007.
58. Pinkiewicz, T. (2012). *Computational techniques for automated tracking and analysis of fish movement in controlled aquatic environments* (PhD thesis). University of Tasmania, Australia.
59. Zhou, C. Zhang, B. Lin, K. & Sun, C. (2017). Near-infrared imaging to quantify the feeding behavior of fish in aquaculture. *CEA 2017, 135*, 233–241.
60. Faro, A., Giordano, D., & Spampinato, C. (2011, December). Adaptive background modeling integrated with luminosity sensors and occlusion processing for reliable vehicle detection. *IEEE Transactions on Intelligent Transportation Systems, 12*(4), 1398–1412.

61. Porikli, F. (2005). Multiplicative background-foreground estimation under uncontrolled illumination using intrinsic images. In IEEE Motion Multi-Workshop.

62. Porikli, F., & Wren, C. (2005, April). Change detection by frequency decomposition: Wave-Back. In IEEE WIAMIS 2005.

63. Bilodeau, G., Jodoin, J., & Saunier, N. (2013, May). Change detection in feature space using local binary similarity patterns. In CRV 2013, pp. 106–112.

64. Ridder, C., Munkelt, O., & Kirchner, H. (1995). Adaptive background estimation and foreground detection using Kalman-filtering. In ICRAM 1995, pp. 193–195.

65. Candes, E., Li, X., Ma, Y., & Wright, J. (2011, May). Robust principal component analysis. *International Journal of ACM, 58*(3).

66. Butler, D., Shridharan, S., & Bove, V. (2003). Real-time adaptive background segmentation. In IEEE ICASSP 2003.

67. Sheikh, Y., & Shah, M. (2005, June). Bayesian object detection in dynamic scenes. In IEEE CVPR 2005.

68. Nazir, S., Newey, S., Irvine, R., Fairhurst, G., & van der Wal, R. (2017). WiseEye: Next generation expandable and programmable camera trap platform for wildlife research. In PLOS.

69. Dickinson, P. (2008). *Monitoring the vulnerable using automated visual surveillance* (PhD thesis). University of Lincoln, UK.

70. Weinstein, B. (2014). Motionmeerkat: Integrating motion video detection and ecological monitoring. In Methods in Ecology and Evolution.

71. Weinstein, B. (2018). Scene-specific convolutional neural networks for video-based biodiversity detection. In Methods in Ecology and Evolution.

72. Boult, T., Micheals, R., Gao, X., & Eckmann, M. (2001, October). Into the woods: Visual surveillance of noncooperative and camouflaged targets in complex outdoor settings. *Proceedings of the IEEE, 89*(10), 1382–1402.

73. Shakeri, M., & Zhang, H. (2017, October). Moving object detection in time-lapse or motion trigger image sequences using low-rank and invariant sparse decomposition. In IEEE ICCV 2017.

74. Yousif, H., Yuan, J., Kays, R., & He, Z. (2017, September). Fast human-animal detection from highly cluttered camera-trap images using joint background modeling and deep learning classification. In IEEE ISCAS 2017, pp. 1–4.

75. Janzen, M., Visser, K., Visscher, D., Mac Leod, I., Vujnovic, D., & Vujnovic, K. (2017). Semi-automated camera trap image processing for the detection of ungulate fence crossing events. In Environmental Monitoring and Assessment, pp. 189–527.

76. Borghgraef, A., Barnich, O., Lapierre, F., Droogenbroeck, M., Philips, W., & Acheroy, M. (2010). An evaluation of pixel-based methods for the detection of floating objects on the sea surface. In EURASIP *Journal on Advances in Signal Processing.*

77. Szpak, Z., & Tapamo, J. (2011). Maritime surveillance: Tracking ships inside a dynamic background using a fast level-set. *Expert Systems with Applications, 38*(6), 6669–6680.

78. Adam, A., Shimshoni, I., & Rivlin, E. (2013, July). Aggregated dynamic background modeling. In *Pattern Analysis and Applications.*

79. Prasad, D., Prasath, C., Rajan, D., & Quek, C. (2017, January). Challenges in video based object detection in maritime scenario using computer vision. *WASET International Journal of Computer, Electrical, Automation, Control and Information Engineering, 11*(1).

80. Prasad, D., Rajan, D., & Quek, C. (2016, November). Video processing from electro-optical sensors for object detection and tracking in maritime environment: A survey. *IEEE Transactions on Intelligent Transportation Systems, 18*(8).

81. Eng, H., Toh, K., Kam, A., Wang, J., & Yau, W. (2003). An automatic drowning detection surveillance system for challenging outdoor pool environments. In IEEE ICCV 2003, pp. 532–539.

82. Eng, H., Toh, K., Kam, A., Wang, J., & Yau, W. (2004). Novel region-based modeling for human detection within highly dynamic aquatic environment. In IEEE CVPR 2004, p. 390.

83. Lei, F., & Zhao, X. (2010, October). Adaptive background estimation of underwater using Kalman-filtering. In CISP 2010.

84. Chan, K. (2013, January). Detection of swimmer using dense optical flow motion map and intensity information. *Machine Vision and Applications, 24*(1), 75–101.

85. Peixoto, N., Cardoso, N., Tavares, A., & Mendes, J. (2012). Motion segmentation object detection in complex aquatic scenes and its surroundings. In INDIN 2012, pp. 162–166.

86. Kavasidis, I., Palazzo, S., & Spampinato, C. (2013). An innovative web-based collaborative platform for video annotation. In Multimedia Tools and Applications, pp. 1–20.

87. Kim, K., Chalidabhongse, T., Harwood, D., & Davis, L. (2004). Background modeling and subtraction by codebook construction. In IEEE ICIP 2004.

88. Radolko, M., & Gutzeit, E. (2015). Video segmentation via a Gaussian switch background model and higher order Markov random fields. In VISAPP 2015.

89. Zivkovic, Z. (2006, January). Efficient adaptive density estimation per image pixel for the task of background subtraction. *Pattern Recognition Letters, 27*(7), 773–780.

90. KaewTraKulPong, P., & Bowden, R. (2001, September). An improved adaptive background mixture model for real-time tracking with shadow detection. In AVBS 2001.

91. Haines, T., & Xiang, T. (2012, October). Background subtraction with Dirichlet processes. In European Conference on Computer Vision, ECCV 2012.

92. Tiefenbacher, P., Hofmann, M., Merget, D., & Rigoll, G. (2014). PID-based regulation of background dynamics for foreground segmentation. In IEEE ICIP 2014, pp. 3282–3286.

93. Maddalena, L., & Petrosino, A. (2008, July). A self-organizing approach to background subtraction for visual surveillance applications. *IEEE T- IP, 17*(7), 1168–1177.

94. Sobral, A., Bouwmans, T., & Zahzah, E. (2015). Double-constrained RPCA based on saliency maps for foreground detection in automated maritime surveillance. In ISBC 2015 Workshop conjunction with AVSS 2015, Karlsruhe, Germany, pp. 1–6.

95. Javed, S., Mahmood, A., Bouwmans, T., & Jung, S. (2017, September). Superpixels based manifold structured sparse RPCA for moving object detection. In BMVC 2017, London, UK.

96. Javed, S., Mahmood, A., Bouwmans, T., & Jung, S. (2017, December). Background-foreground modeling based on spatio-temporal sparse subspace clustering. *IEEE Transactions on Image Processing, 26*(12), pp. 5840–5854.

97. Javed, S., Mahmood, A., Bouwmans, T., & Jung, S. (2016, December). Spatiotemporal low-rank modeling for complex scene background initialization. In IEEE T-CSVT.

98. Bouwmans, T., Javed, S., Zhang, H., Lin, Z., & Otazo, R. (2018). On the applications of robust PCA in image and video processing. In Proceedings of the IEEE, pp. 1427–1457.

99. Braham, M., & Van Droogenbroeck, M. (2016, May). Deep background subtraction with scene-specific convolutional neural networks. In IWSSIP 2016, Bratislava, Slovakia.

100. Minematsu, T., Shimada, A., Uchiyama, H., & Taniguchi, R. (2018). Analytics of deep neural network-based background subtraction. In MDPI Journal of Imaging.

101. Liang, X., Liao, S., Wang, X., Liu, W., Chen, Y., & Li, S. (2018 July). Deep background subtraction with guided learning. In IEEE ICME 2018, San Diego, CA.

102. Sultana, M. Mahmood, A. Javed, S. & Jung, S. (2019). Unsupervised deep context prediction for background estimation and foreground segmentation. *Machine Vision and Applications, 30*(3), 375–395.

103. Laugraud, B., Pierard, S., & Van Droogenbroeck, M. (2018). LaBGen-P-Semantic: A first step for leveraging semantic segmentation in background generation. *MDPI Journal of Imaging, 4*(7), Art. 86.

104. Laugraud, B., Pierard, S., & Van Droogenbroeck, M. (2017). LaBGen: A method based on motion detection for generating the background of a scene. In Pattern Recognition Letters.

105. Laugraud, B., Pierard, S., & Van Droogenbroeck, M. (2017). LaBGen-P: A pixel-level stationary background generation method based on LaBGen. In IAPR ICPR 2016.

106. Javed, S., Bouwmans, T., & Jung, S. (2017). SBMI-LTD: Stationary background model initialization based on low-rank tensor decomposition. In ACM SAC 2017.

107. Sobral, A., Bouwmans, T., & Zahzah, E. (2015, September). Comparison of matrix completion algorithms for background initialization in videos. In ICIAP 2015, Genoa, Italy.

108. Zhang, Y., Li, X., Zhang, Z., Wu, F., & Zhao, L. (2015, June). Deep learning driven blockwise moving object detection with binary scene modeling. In Neurocomputing.
109. Shafiee, M., Siva, P., Fieguth, P., & Wong, A. (2016, June). Embedded motion detection via neural response mixture background modeling. In IEEE CVPR 2016.
110. Shafiee, M., Siva, P., Fieguth, P., & Wong, A. (2017, June). Real-time embedded motion detection via neural response mixture modeling. In Journal of Signal Processing Systems.
111. Nguyen, T., Pham, C., Ha, S., & Jeon, J. (2018, January). Change detection by training a triplet network for motion feature extraction. In IEEE T-CSVT.
112. Sheikh, Y., Javed, O., & Kanade, T. (2009, October). Background subtraction for freely moving cameras. In IEEE ICCV 2009, pp. 1219–1225.
113. Elqursh, A., & Elgammal, A. (2012). Online moving camera background subtraction. In European Conference on Computer Vision, ECCV 2012.
114. Sugaya, Y., & Kanatani, K. (2004). Extracting moving objects from a moving camera video sequence. In Symposium on Sensing via Image Information, SSII 2004, pp. 279–284.
115. Taneja, A., Ballan, L., & Pollefeys, M. (2010). Modeling dynamic scenes recorded with freely. In Asian Conference on Computer Vision, ACCV 2010.

6

Internet of Things and Remote Sensing

Pradeep K. Garg
Civil Engineering Department, Indian Institute of Technology, Roorkee, India

CONTENTS

6.1 Introduction

The Internet of Things (IoT) consists of technologies that connect the real world of physical objects [1]. The IoT is the next phase of the information revolution, which involves billions of smart devices and sensors interconnected to provide speedy information and real-time data sharing. The IoT extends the "anywhere, anyhow, anytime" computing and networking paradigm to "anything, anyone, and any service." The IoT can be viewed as a networks of networks. It can be viewed today as billions of connections that will encompass every aspect of our lives.

The vision of IoT involves seamless integration of physical devices to the Internet/web by means of well-understood, accepted, and used protocols, technologies, and programming/description languages for efficient human-to-machine or machine-to-machine (M2M) communication [2]. The idea behind the IoT is to make things intelligent, programmable, and more capable of interaction with humans. It not only enhances the comforts of our lives, but it also gives us more control by simplifying our routine work and tasks. Deploying IoT applications provides several advantages. The principal benefit is that it helps organizations make real-time, automated, intelligent decisions that are driven by analytics. Thus, the potential of the IoT market is huge.

Cloud computing provides the capability of delivering ubiquitous services on-demand, as per requirements of applications [3, 4]. Cloud computing has been adopted in remote sensing-based applications, such as the Mat-Su project for flood assessment. Smart cities effectively use cloud computing applications such as Apache Spark and Hadoop to analyze huge data collected by IoT devices.

Artificial intelligence (AI) plays an important role in IoT-based applications. IoT software platforms offer integrated AI capabilities, such as machine learning-based analytics. While IoT connects machines and makes use of the data generated from those machines, AI simulates the intelligent behavior of all such machines [5]. The field of AI is older than big data or the IoT. Big data, IoT, and AI are already individually very powerful and innovative technologies; combining them creates further synergies. Big data forms the center, whereas IoT and AI provide support at different stages in the process [2]. The IoT allows capturing much additional data. There have been and there are still other traditional sources for big data, but the IoT has enormous potential to provide much more data, accurately and in real-time. AI allows analyzing data in new ways. There have been and there are still other traditional methods of analyzing big data, but AI has enormous potential to analyze much more data to deliver actionable information in real-time.

6.2 Geospatial Big Data

Big data is the aggregated data from lots of little data, and any error/anomaly in little data can cause problems with the big data. A large variety of geospatial data is available, which is no longer limited to simply two-dimensional (2D) images. It now includes three-dimensional (3D) data in the form of locational coordinates obtained from GPS and light detection and ranging (LiDAR) technology [6]. Geospatial data can be collected from aircraft, satellites, UAVs, mobile and stationary cameras and scanners, positioning devices, and terrain models. Most IoT data are often geospatial, but there has been little interaction between the IoT and geodata. Geodata collection and analysis are therefore essential to provide timely and reliable data for IoT applications [7, 8].

Requirements of geospatial data have changed today from 2D drawings and maps to facilitate terrain evaluation and visualization. Three-dimensional models can be generated from geospatial data as required in many applications. By using multiple geospatial datasets, four-dimensional (4D) models can be developed through software that allows users to view changes/growth of objects over time. Such models can be used to measure changes in infrastructure, traffic, earthworks, and agriculture. Multi-time images allow users to carry out time-based analyses. Comprehensive analysis of time-series geospatial information enables us to examine not only the changes but also the relationships between the factors and changes. Effective applications, such as house, transportation, gas, power, and agriculture, depend on large data collection that can be acquired using geospatial information for 3D and 4D analyses [9].

Geospatial big data provides detail and contextual information that is useful in many applications. The use of IoT in a geospatial context can also be associated with time-related data that record continuous measurements allowing us to study phenomena such as landslides [10], floods, and the activity and behavior of wild animals [11]. Historical time series-based observations are essential to analyze for forecasting disasters such as flooding [12] and forest fires [13].

These data are so large and complex that advanced tools and skills are required to manage, process, and analyze them. Specialized solutions, such as automated 3D modeling and feature-recognition software, may further increase the value of utilization of big data by extracting specific information from large images and points cloud data. In most applications, real-time data are collected from wireless communications connected to the cloud.

The speed, ease, and flexibility of this technology help to increase the use of real-time global navigation satellite system (GNSS) positioning [12]. Decision makers require real-time information that relies on technologies such as communications systems, sensors, and application-based software.

6.2.1 Using Sensors and IoT

IoT devices are used in many applications including sensors, radio frequency identification (RFID) tags, barcodes and quick response (QR) codes [1], mobile phones and their embedded sensors (i.e., cameras, microphones) [14], and GPS in studies dealing with participatory sensing and crowdsourcing [15]. Several other sensors are also used, such as water sensors; pollution sensors; infrared (IR) cameras; temperature, humidity, and pressure sensors [16]; meteorological and hydrological sensors [12, 17]; water column height sensors [18]; and animal collar tags [19]. Some applications supplement their data from IR cameras and mobile phone-based crowdsourcing with web-based sources, such as online news, road data, medical facility information, disaster-relief scenarios [19], and geo-tagged photos [20].

Based on past worldwide studies, Kamilaris and Ostermann [21] identified six different methods where IoT-based sensors and their measurements can be used to generalize larger-scale outputs from a smaller number of point measurements. These IoT-based methods include participatory sensing, vehicular networks and transportation systems, fixed IoT sensor installations, satellite imagery, ground sensor sampling, and web-based IoT datasets. Some sensors have some limitations. For example, GNSS receivers provide low accuracy due to weak signals received in high-density urban areas; therefore, a tag location-based technique is recommended to improve the accuracy from such weak signals.

6.2.2 Using GPS and IoT

Internet-enabled things (i.e., sensors, mobile phones, cameras) are normally equipped with sensors that can measure temperature, humidity, velocity, time, radiation, electromagnetism, and so on. Multisensory mobile phones have advanced sensing capabilities, such as measuring proximity, acceleration, and location; recording audio/noise; sensing electromagnetism; and capturing images and videos [20, 22]. An important characteristic of the measurements performed by these Internet-enabled things is the tagging of geographic location of each measurement (i.e., latitude and longitude coordinates), which is quite useful in many applications.

Figure 6.1 presents various applications of IoT in mobile computing, along with the sensors and IoT devices [23]. Location-based sensing is important in applications having interactions between humans and physical things, and also those having social interactions with people [22]. The application domains of IoT and location-based projects are increasing, and now include smart homes [7], urban environments [24], and smart agriculture [25]. For location-based applications, understanding geographic information in relation to the local environment is crucial [26], particularly in mobile phone computing [27]. Location-based sensing is frequently used in both outdoor and indoor positioning [28].

6.2.3 Using UAV and IoT

UAVs, or drones, have become widespread in all aspects of life, and their popularity has increased immensely over the last 5 years. However, the history of drones can be traced

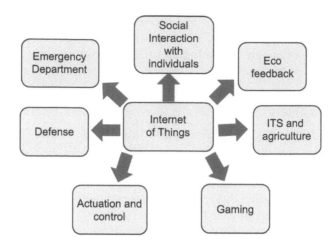

FIGURE 6.1
The location-based smartphone Internet of Things applications [23].

back more than 100 years as a fixed-wing military development during World War I. The nickname *drone* comes from the worker bee. Drones, or UAVs, are very useful in many applications as they provide high-resolution images and laser data, and thus save some of the time and effort required to collect field data. In practice, two types of drones are used for IoT applications: ground-based and aerial-based [29].

Initially developed for military operations, UAVs have become popular in recent years for civilian customers. UAV images belong to the big data category "sensing data." UAV imagery is useful to assess large and not easily accessible areas [2]. For example, the UAV has been used in a study in Australia that takes frequent measurements at mine sites to capture aerial images. These images are collected at regular intervals at much lower cost than traditional aircraft photogrammetric images. An additional advantage is that the UAV can collect data of inaccessible or dangerous sites. Without disrupting normal mine operations or sending field surveyors into hazardous areas, the UAV can provide useful images. These images are processed to produce 3D models for viewing the terrain or for volume computations.

For agricultural applications, drones can assist in soil analysis (with 3D maps for seed planting predictions), planting (by providing information on nutrients), crop monitoring (by providing time-series satellite images), irrigating the field (having sensors to predict dry areas), and crop health (taking crop scans). In other words, drones can manage the full cycle of crops. Drones can provide details on plant counting, yield prediction, health indices, height, the presence of chemicals in plants and soil, drainage mapping, and various other data. With the pictures and aerial maps provided by drones, farmers can immediately understand which crops need urgent attention. In addition to these advantages, the agricultural innovation of drones promotes better care of crops in general, such as evaluation of their health state, irrigation, monitoring of progress, spraying, and planting [30].

6.2.4 Using Satellite Images and IoT

Satellite images are further classical sources of big data. Satellite images are not new, but the number of satellite images keeps rising, as the coverage is more comprehensive and the resolution of the images gets sharper to outline more features. Satellites are conducting "remote sensing." That is, sensors are gathering data for objects without being in physical contact with the objects. Satellite imagery is very useful to assess large areas, for example, agricultural land, land cover and land use in general, and also the condition of ecosystems and natural disaster impacts (e.g., after earthquakes and flooding [2]). The quality of satellite images is diminished when the areas of interest are covered by clouds.

Satellite data are important in many applications, such as smart city, disaster monitoring, climate prediction, agriculture management, floods, urban sprawl, and remote surveillance. High-resolution satellite images from IKONOS, QuickBird, or WorldView are used by insurance and financial companies for detailed crop mapping or resource mapping to provide compensation to their customers.

In addition to geo-located information, a variety of data from different sources can be used for enhancing the geospatial analysis, such as mobile phone-based crowdsourcing [15] or satellite-based imagery [31]. Digital map layers, including thematic maps [32], soil maps [8], and digital elevation models (DEMs), are vital in many land-based studies [33]. The nature of data used in these studies may include text, images, sound, video, and location-based data.

The big remote-sensing data, along with the data collected from connected IoT devices (e.g., smart watches, vehicles), offers several new avenues for high-performance computing platforms, such as clusters and supercomputers, which have already been tested for data processing, analysis, and knowledge discovery. This, combined with IoT data, can dramatically enhance the process of analyzing the big data, and derive useful inferences for the applications.

Remote sensing provides the capability for industries to gain competitive intelligence. For example, Orbital Insight uses AI and big data analytical technologies remote-sensing imagery to deliver action plans for investors, governments, and environmentalists [34]. Orbital Insight explores how remote-sensing images can be used as effective tools for decision makers across all industries. The company aggregates millions of satellite images, analyzes them with machine vision, and extracts useful information for decision makers. This technology can also be used to detect the number of cars in parking lots to determine quarterly and yearly sales for major retailers, to track roads being built in forests to predict illegal deforestation, and so on. For example, Walmart cannot install a traffic counter inside a store, but high-resolution satellite data can show Walmart how many cars were parked in the parking lots of stores, which has a direct correlation to the company's sales.

6.3 Connection between Geospatial Analysis and IoT

With the development of IoT and big geospatial data, isolated companies are now very much interconnected. Industries are heavily dependent on real-time geospatial data, be it enterprise-scale operations or consumer applications on a smartphone. The required data and information are gathered from every possible source, analyzed, and converted into actionable intelligence called *geospatial intelligence* [21, 35]. Geospatial intelligence is now carried out with the IoT because it provides the most up-to-date spatio-temporal information needed by businesses for day-to-day reliable decision-making.

Spatial analytics in combination with the IoT can be more than just features on a map, as it deals with the data relating to the position, size, or shape of objects in 2D or 3D space. Geospatial data offers great potential for better understanding, modeling, monitoring, and visualizing the objects, using IoT as an important tool [9]. The applications of spatial analytics in the IoT are challenging because (i) large amounts of data must be extracted, stored, and processed; (ii) many sources provide heterogeneous data that are to be integrated; (iii) many applications require real-time data processing; and (iv) there is a greater need for creating visualization models.

There is increasing awareness of the mapping and analytical capabilities of geographic information system (GIS) mapping technology. As GISs are used more frequently, their applications are being combined with IoT initiatives in business and other areas. Interactive GIS-based mapping along with IoT data can significantly improve the understanding of large data. GIS applications are being developed in business to gain a better understanding of geographical areas and patterns [31]. Companies are already analyzing geography in numerous ways to improve their efficiency and functionality. These companies are gaining a better understanding to build new facilities, store resources, minimize waste, and lay down power lines or electrical cables.

GIS can effectively carry out distance and proximity/neighborhood analysis, which is a requirement in applications such as tourism [14] or location-based marketing [23], in which tourists and citizens, based on their current locations, could be offered marketing services. Some examples include specific offers available in malls, hotels, shops, or sightseeing locations. Other application areas include mobile dating (geo-social services), such as Grindr [36] and OkCupid [37], as well as gaming, such as the popular Pokemon Go [38]. Distance and proximity/neighborhood analysis is also helpful for travel by roads, rails, and other public transit; fuel requirements in vehicles; and management of infrastructures [18].

Spatial features can be utilized, for example, to visualize the locations of objects (e.g., a delivery van) at a given time and calculate the distance. It is possible to track objects in real-time, if the location information is stored with time. This data would generate new applications, such as shortest-path analysis for travel. In the next step, advanced analytics and machine learning tools can be employed to derive the patterns. Finally, the information would help in supporting decisions (e.g., optimization of routes), or even predicting behaviors [6].

To support the functionality given in Figure 6.1, spatial analytics comprises four steps:

1. **Collection of spatial data:** The data are collected from various data sources, such as web services, satellite images, GPS, UAV, open data, sensors, or other systems. This could be either one-time data, or data collected on a regular basis or in near real-time with streaming technology.

2. **Storage and management of spatial data:** Spatial data are to be stored in a warehouse, which serves as an integrated data repository. The spatial data warehouses are usually extended with specific structures and models optimized for storing and processing locational data [3]. Moreover, the warehouses frequently encompass big data and NoSQL (not only SQL) technologies to enable heavy-write or heavy-read scenarios [6].

3. **Analysis and modeling of spatial data:** Large amounts of spatial and nonspatial datasets are combined to support complex geo-algorithms. Extensive research is therefore required for efficient implementation of spatial data and the exploitation of upcoming technologies, such as graph databases.

4. **Visualization and integration of spatial data:** The processed data are presented in an appropriate form to facilitate and support decision-making. For this, analytics and dashboard tools allow users to visualize data on a map or more sophisticated visualizations, or to do visualization on spatial data.

TABLE 6.1

Relationship between IoT Research Areas and Geospatial Analysis Methods Used

IoT Area	Geometric Measures	Data Mining	Basic Analytical Operations	Basic Analytical Methods	Network Analysis	Surface Analysis and Geostatistics
Disaster monitoring	X	-	X	-	-	X
Wildlife monitoring	-	X	-	X	X	X
Agriculture	X	-	-	X	X	X
Environmental monitoring	-	-	X	X	X	X
Biodiversity	-	-	-	X	X	-

Kamilaris and Ostermann [21] assessed 63 research papers based on the IoT, where geospatial analysis is used in environmental informatics. Six different geospatial analysis methods are identified by them, and presented together with 26 relevant IoT initiatives. Table 6.1 lists each IoT research area, as well as the geospatial analytical method employed. For example, the geometric measure that deals with distances and proximity, as well as adjacency and connectivity of points is the basic function of geospatial analysis. Because of their quick derivation of these data, geometric measures have several applications in mobile location-based services [23], such as transportation, traffic, health, and agriculture. Also used is spatial data mining, which finds patterns and extracts information from large geospatial data [32].

Further, Kamilaris and Ostermann analyzed the types of IoT devices employed, their deployment status and data transmission standards, data types used, and reliability of measurements. This study also provides some indications of how IoT, together with geospatial analysis, is being used currently in the domain of environmental research. In this study, IoT applications are examined that employ various geospatial analysis techniques and analytic operations and focus on location. A more general survey on IoT and geospatial analysis is given by Brunkard [5].

6.4 Conclusion

Hardware advances in space have directly led to software advancements. The challenge is to know how to apply the right tools at the right time to the right problem at hand [1]. Geospatial big data may be complex, but its analysis is extremely valuable to its users [32]. The IoT and geospatial data are interconnected, as users come to rely more on geospatial data for development of applications. The number of connected sensors is increasing very quickly. At present, there are billions of connected devices around the world; for example, the utility market uses smart sensors. Smartphones and wearable technology will soon become major economy drivers in the marketplace, as these IoT-based devices will produce data that are required for intelligent decision-making.

Location-based sensors are one of the most employed sensors in the IoT, as they allow locating and tracking the objects accurately. Spatial analytics techniques heavily depend on the reliability of available data; therefore, the data become more valuable when physical locations are captured more precisely in real-time. GIS integration with the IoT may lead to entirely new practices that are safer and more efficient [31]. Spatial data obtained from remote-sensing images could be used to suggest actions generated through map analysis. These types of results demonstrate the potential of GIS and IoT integration to change the working of industries and create new initiatives. This is where the IoT plays a prime role and will be a major driver for spatial analytics. Geospatial analysis using high-resolution satellite images offers a great potential for better modeling, monitoring, and visualizing of processes, using IoT as an important tool [32].

Cloud-based positioning services are used in several applications, including surveying, construction, traffic, and agriculture. For example, even in remote areas, cloud positioning and web interfaces can be used to monitor a construction site for delivering real-time information to its stakeholders [4]. In the current scenario, clouds and big data processing are not designed to deploy remote-sensing data. Current techniques do not provide the flexibility to fuse remote-sensing data with data generated locally or via connected IoT devices. For example, optimizing the use of remote-sensing data combined with data from GIS and IoT devices can lead to several applications, such as sustainable agriculture and smart cities.

Satellite technology plays a key role in driving the growth momentum behind the IoT and interconnected devices [6]. Connectivity in most remote locations, and even in the air and at sea, enables IoT applications across diverse market segments. Satellites provide ubiquitous coverage for a truly connected world. Before data from IoT devices can be used for applications, the data must be confirmed to be reliable with accurate measurements. Accuracy issues may be more relevant while using satellite images, due to their spatial resolutions [9]. As more and more satellites are launched into space in the future, persistent surveillance of the globe may help us to build a more sustainable environment.

References

1. Atzori, L., Iera, A., & Morabito, G. (2010). The Internet of Things: a survey. *Computer Networks, 54*(15), 2787–2805.
2. Ziesche, S. (2017, May 31-June 2). Innovative big data approaches for capturing and analyzing data to monitor and achieve the SDGs. Paper presented at Forum on Innovative Data Approaches to SDGs, Incheon, Republic of Korea.
3. https://cables24.com/en/blog/6_brief-history-of-the-internet-of-things
4. Cisco. (2015). Fog computing and the Internet of Things: Extend the cloud to where the things are. https://www.cisco.com/c/dam/en_us/solutions/trends/iot/docs/computing-overview.pdf
5. Brunkard, P. (2018). The future of IoT is AI. https://www.techuk.org/insights/opinions/item/13827-the-future-of-iot-is-ai
6. Ereth, J. (2018). Geospatial analytics in the Internet of Things. https://www.eckerson.com/articles/geospatial-analytics-in-the-internet-of-things
7. Kamilaris, A., Pitsillides, A., & Yiallouros, M. (2013). Building energy-aware smart homes using web technologies. *Journal of Ambient Intelligence and Smart Environments, 5*(2), 161–186.
8. Kamilaris, A. (2017, September 13-15). Estimating the environmental impact of agriculture by means of geospatial and Big Data analysis: The case of Catalonia. In Proceedings of the EnviroInfo, Luxembourg.

9. Kamilaris, A., & Ostermann, K. O. (2018). Geospatial analysis and the Internet of Things. *ISPRS International Journal of Geo-Information, 7*(268), 1–22.

10. Benoit, L., Briole, P., Martin, O., Thom, C., Malet, J. P., & Ulrich, P. (2015). Monitoring landslide displacements with the Geocube wireless network of low-cost GPS. *Engineering Geology, 195*(10), 111–121.

11. Ayele, E., Das, K., Meratnia, N., & Havinga, P. (2018). Leveraging BLE and LoRa in IoT network for wildlife monitoring system (WMS). In Proceedings of the IEEE 4th World Forum on Internet of Things, Singapore.

12. Berger, H. (1991). Flood forecasting for the river Meuse. *Hydrol. Water Manag. Large River Basins, 201,* 317–328.

13. Puri, K., Areendran, G., Raj, K., & Mazumdar, S. (2011). Forest fire risk assessment in parts of Northeast India using geospatial tools. *Journal of Forestry Research, 22*(4), 641.

14. Van Setten, M., Pokraev, S., & Koolwaaij, J. (2004). Context-aware recommendations in the mobile tourist application COMPASS. *Adaptive hypermedia and adaptive Web-based systems* (pp. 235–244). Berlin/Heidelberg, Germany: Springer.

15. Liu, J., Shen, H., & Zhang, X. (2016). A survey of mobile crowdsensing techniques: A critical component for the internet of things. In Proceedings of the 25th International Conference on Computer Communication and Networks (ICCCN), Waikoloa, HI, pp. 1–6.

16. Kotsev, A., Schade, S., Craglia, M., Gerboles, M., Spinelle, L., & Signorini, M. (2016). Next generation air quality platform: Openness and interoperability for the Internet of things. *Sensors, 16*(3), 403.

17. Hay, S., & Lennon, J. (1999). Deriving meteorological variables across Africa for the study and control of vector-borne disease: A comparison of remote sensing and spatial interpolation of climate. *Tropical Medicine & International Health, 4*(1), 58–71.

18. Perumal, T., Sulaiman, M., & Leong, C. (2015). Internet of Things (IoT) enabled water monitoring system. In Proceedings of the IEEE 4th Global Conference on Consumer Electronics (GCCE), Osaka, Japan, pp. 86–87.

19. Zook, M., Graham, M., Shelton, T., & Gorman, S. (2012). Volunteered geographic information and crowdsourcing disaster relief: A case study of the Haitian earthquake. World Medical & Health Policy, 2, 7–33.

20. Kisilevich, S., Mansmann, F., & Keim, D. (2010). P-DBSCAN: A density based clustering algorithm for exploration and analysis of attractive areas using collections of geo-tagged photos. In Proceedings of 1st International Conference and Exhibition on Computing for Geospatial Research & Application, Bethesda, MD, p. 38.

21. Kamilaris, A., & Ostermann, F. (2018). Geospatial analysis and Internet of things in environmental informatics. https://arxiv.org/abs/1808.01895

22. Küpper, A. *Location-based services: Fundamentals and operation.* (2005). Hoboken, NJ: John Wiley & Sons, Ltd.

23. Rao, B., & Minakakis, L. (2003). Evolution of mobile location-based services. *Communications of the ACM, 46*(12), 61–65.

24. Kamilaris, A. Pitsillides, A., Prenafeta-Boldu, F., & Ali, M. I. (2017). A web of things based ecosystem for urban computing—towards smarter cities. In Proceedings of the 24th International Conference on Telecommunications (ICT), Limassol, Cyprus.

25. Kamilaris, A., Gao, F., Prenafeta-Boldu, F., & Ali, M. I. (2016, December 12-14). IoT: A semantic framework for Internet of Things-enabled smart farming applications. In Proceedings of the IEEE World Forum on Internet of Things (WF-IoT), Reston, VA.

26. Wilde, E., & Kofahl, M. (2008, April 22). The locative web. In Proceedings of the First International Workshop on Location and the Web (LOCWEB '08), Beijing, China, pp. 1–8.

27. Kamilaris, A., & Pitsillides, A. (2016). Mobile phone computing and the Internet of Things: A survey. *IEEE Internet Things (IoT) Journal, 3*(6), 885–898.

28. Zeimpekis, V., Giaglis, G., & Lekakos, G. (2002). A taxonomy of indoor and outdoor positioning techniques for mobile location services. *SIGecom Exchanges.* doi:10.1145/844351.844355

29. Lagkas, T., Argyriou, V., Bibi, S., & Sarigiannidis, P. (2018). UAV IoT framework views and challenges: Towards protecting drones as "things." *Sensors, 18*(11), E4015. doi:10.3390/s18114015

30. Ahn, T., Seok, J., Lee, I., & Han, J. (2018). Reliable flying IoT networks for UAV disaster rescue operations. *Mobile Information Systems*, Article ID 2572460. https://doi.org/10.1155/2018/2572460

31. Esri. How to harness the full potential of IoT data. (2018). http://www.esri.com/iot

32. Van der Zee, E., & Scholten, H. (2014). Spatial dimensions of big data: Application of geographical concepts and spatial technology to the internet of things. In *Big data and Internet of things: a roadmap for smart environments* (pp. 137–168). Berlin, Germany: Springer.

33. Bisio, R. How spatial data adds value to the Internet of things. (2018). http://internetofthingsagenda.techtarget.com/blog/IoT-Agenda/How-spatial-data-adds-value-to-the-internet-of-things

34. https://orbitalinsight.com/

35. Shekhar, S., Zhang, P., Huang, Y., & Vatsavai, R. R. (2003). *Spatial data mining*. State College, PA: Citeseer.

36. https://www.grindr.com/

37. https://www.okcupid.com/

38. Niantic, Inc. https://pokemongolive.com/es/

7

Internet of Things and Cloud Computing

Umang Kant[1], Mayank Singh[2], and Viranjay M. Srivastava[2]

[1]*Krishna Engineering College, Ghaziabad, Uttar Pradesh, India*
[2]*University of KwaZulu-Natal, Durban, South Africa*

CONTENTS

7.1 Introduction

The IoT is the current big thing with respect to the Internet. It is a network of physical objects that use sensors and application programming interfaces (APIs) to connect and analyze data over the Internet. To connect devices to the Internet, IoT depends on a complete host of technologies, such as big data, predictive analytics, artificial intelligence (AI) and machine learning, cloud computing, and RFID [1]. Cloud-based IoT platforms and architecture connect the real world with the virtual world by helping companies collect device data, link the devices with back-end systems, build and run IoT applications, and manage IoT devices with connectivity and security. Predictive analytics and big data enter the picture when enormous amounts of data are generated by the smart devices; these data must be analyzed and controlled in real-time. Machine learning is used when actions without human intervention are required. A subset of IoT used in the field of manufacturing is the Industrial Internet of Things (IIoT), which is also known as Industry Internet or

Industry 4.0 [2]. IIoT requires M2M technology to support remote monitoring, predictive maintenance, telemarketing, and other tasks.

IoT provides attractive opportunities for companies to develop novel services enabled with real-time sensors. Many companies use IoT and cloud computing-enabled applications to automate manufacturing and business processes, remotely manage operations, optimize supply chains, and conserve resources. This technology enables users and employees of multinational companies to automate routine and repetitive tasks, improve decision-making, and accelerate communication, and also makes it simple for users to manage day-to-day activities. From incorporating customer care with actual product performance and usage to delivering products and services, IoT offers many ways to provide appealing and user-friendly customer experiences across the globe. Apart from the benefits of IoT discussed above, one of its major benefits is efficiency [3].

The aim of IoT is to achieve a world of interconnected devices that are smart enough to share information with other devices, users, and cloud-based applications. This concept of connectivity goes beyond our smartphones, laptops, and tabs to include smart homes, smart cars, smart wearable devices, smart cities, smart healthcare organizations, and smart banking. This aim points towards a complete automated life, and hence a complete automated world. According to a report by Gartner, IoT is estimated to have around 20.6 billion connected devices by the end of 2020 [4]. In 2015–2016, HP conducted a small survey about the rise of connected devices over the years, which provided some astonishing results: In the year 2013, the number of connected devices was 9 billion, by the year 2025 it will be 1 trillion!

Smart devices improve the quality of personal and social lives of individuals and of a society as a whole with a broad spectrum of applications. KRK Research conducted a survey in UK, US, Japan, and Germany, wherein the results depict the devices customers are most likely to use in the coming years [5]. It is also discussed as per the report generated by Cisco that the IoT will have a huge impact on the world economy, as it will help generate approximately $14.4 trillion in value across all industries combined.

7.2 Working of IoT

IoT facilitates human-machine interactions (HMIs) and M2M interactions by connecting multiple devices to the Internet simultaneously. As shown in Figure 7.1, IoT has four fundamental components [6]: (1) sensors devices, (2) data processing, (3) user interface, and (4) connectivity.

These components explain how IoT works:

1. **Sensor Devices:** A device can have a single sensor or multiple sensors. Device sensors collect all microscopic data and details from the respective environment. These collected data from various devices have many levels of complexities depending upon the application. This is the most basic step to collecting all the data.

2. **Connectivity Components:** Cloud computing comes in this phase. Here, the collected data are directed towards the cloud infrastructure with the help of a medium. The medium of communication, which ranges from Wi-Fi, Bluetooth,

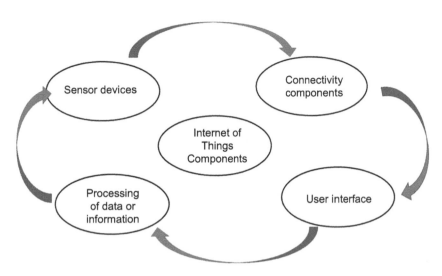

FIGURE 7.1
Major components of Internet of Things.

cellular, and satellite networks, connects the cloud with the device sensors. The medium is selected based on the application specifications and trade-offs between bandwidth, range, and power consumption. Hence, the selection of the connection medium for communication is of key importance to get the best results.

3. **Processing of Data:** Processing enters the picture once the data are collected and transferred to the cloud. There are many application-based dedicated software programs for data processing. Processing ranges from simple (using simple algorithms to check the temperatures on devices such as air-conditioning and washing machines) to very complex (identifying objects such as intruder detection in homes using computer vision).

4. **User Interface:** The processed information is made accessible to the users during some interface on the IoT system (notifications through texts or emails, triggering alarms such as in the case of intruder detection in the home). Users may make a decision manually (e.g., decreasing the treadmill speed when alerted to increased heart rate), or decisions can be automatically made by establishing and implementing predefined protocols (e.g., air-conditioning temperature control).

An IoT platform is required to provide an environment to devices and objects with built-in sensors. This platform connects these devices and objects, and integrates the data provided by them. It then applies analytics to share the related information with applications developed to address specific user needs. These powerful IoT platforms identify and suggest what information should be considered and what can be safety disregarded [7]. This information can be used to detect patterns (such as profit, loss, increment, decrement), detect problems before they have occurred and suggest possible remedies, and make recommendations and suggestions for making decisions.

For example, in a retail business, using an IoT platform, the owner can:

1. use sensors to detect which areas in the store are the most popular, i.e., where the customers like to linger more,

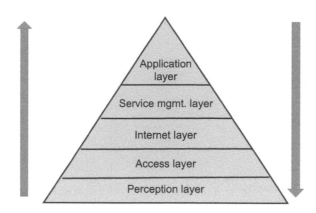

FIGURE 7.2
The Internet of Things environment.

2. identify patterns of customers regarding the choice of products while shopping,

3. analyze the available sales data to identify which products are selling more and fast, and

4. match sales data with supply data, so that popular items do not go out of stock.

The IoT environment is composed of various layers that enable the working of IoT platforms discussed above. As depicted in Figure 7.2, the functionality of each layer is dependent upon the previous-level layer. These layers are: (1) perception layer, (2) access layer, (3) Internet layer, (4) service management layer, and (5) application layer [8].

1. **Perception Layer:** This layer uses sensors attached in the devices present and connected in that environment to capture environment-monitoring information (such as humidity, temperature, and light). This layer requires the use of wireless sensor network technology to form an autonomous network to manage the collaborative work to extract useful information [8].

2. **Access Layer:** The primary function of the access layer is to transfer the captured information from the perception layer to the Internet through public communication networks (such as WLAN, GSM network, satellite network, and other network infrastructure).

3. **Internet Layer:** The Internet layer supports cloud computing as discussed above. It integrates the information resources within the public communication networks into a large smart network (cloud network) that interconnects the devices with sensors to provide an efficient and reliable infrastructure platform for upper service management and application monitoring.

4. **Service Management Layer:** This layer provides real-time management and control of the captured information, and provides a user interface to the respective application. Basically, this layer is responsible for providing service to the users.

5. **Application Layer:** The application layer integrates the system functioning with industry application services to provide service to the users.

7.3 Technologies Used in IoT

Technologies making IoT a reality are broadly categorized into (a) building blocks of IoT, (b) RFID, (c) cloud computing, and (d) M2M communication. This section consists of brief descriptions of each category [9].

a. **Building Blocks of IoT:** The pillars that make IoT sustainable are (i) sensors/sensor network, (ii) embedded processing, and (iii) communication. These building blocks bring intelligence into things. These are explained in brief:

Sensor Network in IoT Key Technology: As discussed earlier, IoT consists of a collaborative environment, which in turn is composed of interconnected devices with sensors. Because these devices are interconnected and collaborative, they form a sensor network. Hence, IoT is composed of a sensor network along with other components in the environment. The sensor network technologies primarily comprise data acquisition, signal processing, protocols, device management, security, network access, intelligent information processing and integration, and other aspects [8].

Data Acquisition Technology: Being the foundation IoT, data acquisition is the fundamental requirement of IoT and is associated with the links and interaction between devices (objects) and users. Data acquisition devices usually have embedded microcontrollers (MCUs) [5]. Sensor technology is the vital technology for data acquisition. Factors such as temperature, humidity, pressure, illumination, and obstacles can be detected by sensors of specific-purpose applications. Hence, sensors are the key components of data acquisition and in turn IoT data acquisition technology includes sensor technology and embedded systems technology.

Signal Processing Technology: Signal processing or signal analysis technologies comprise many variations, such as classification, feature comparison, and so on. These signal analysis technologies map feature signals to a class of physical events. Different functionalities of signal processing are signal interference, signal separation, and signal filtering [8, 9]. Intelligent signal processing is capable of treating the original signal by making the required application-specific modifications to obtain the required results. First, various attributes of the original signals are recorded. Second, signal extraction technology filters the useful signals to separate them from the original signals. Third, conditioning to improve signal-to-noise ratio is applied. Factors to keep in mind are monitoring and managing the data traffic, energy consumption, and network costs within the network.

Network Access: The access to the network is provided by connecting terminal sensor networks with various network access technologies available in the market, such as WLAN, WPAN, SCDMA, GSM, and so on. Network access of IoT is achieved by gateway and other resources available depending upon the application in service.

Information Processing and Integration: Data or information processing has been discussed in various sections in this chapter. It is a basic step and requirement to support IoT. Information integration is crucial, and it is a method of intelligent information processing. It involves multilevel, multidimensional processing that uses the data from multiple data sources or multiple sensors or multiple sensor networks. Evidently, it achieves higher accuracy than a single sensor by managing the data from different nodes or sources in a combined architectural pool. This technology is required in perception, access, Internet, and application layers of the IoT environment (discussed in the previous section) [10].

Interaction and Collaboration: Interaction is an integral component of IoT application systems. It manages the physical signals of perception and makes decisions based on the situations and events of the outside world. It also monitors and adjusts the system's behavior to achieve intelligent interaction. Meanwhile, for IoT research, the focus is on collaborative sensing technology. The data of multiple (collaborating) sensors are combined together to achieve co-perception.

b. **Radio Frequency Identification (RFID) Technology:** RFID belongs to a set of automatic identification data capture (AIDC) technologies. The AIDC technologies automatically recognize the objects in the environment, analyze and collect data about them, and store the captured data into the database or cloud without human intervention. RFID refers to a technology where readers capture digital data encoded in RFID tags or smart labels via radio waves. RFID takes inspiration from barcoding. In barcoding, the data from a label or tag are captured by a dedicated optical-scanner device and stored in a database. These days in smart applications, RFID is preferred over barcoding because barcodes must be aligned with the optical-scanner device whereas RFID tags or labels can be read outside the line of sight [9].

The RFID system consists of (1) the RFID tag (smart label), (2) the RFID reader (interrogator), and (3) an antenna. The integrated circuit in the RFID system consists of the antenna that is used to transmit the captured data of the identified object to the RFID reader, which in turn converts the radio waves to functional or usable data. Finally, the information collected by RFID tags is transmitted to a host computer server through a communication medium. This information is then stored in a database, and is ready for further analysis. Simplified working of RFID technology is depicted in Figure 7.3. RFID technology is being used in many applications making it suitable for IoT applications as well [11, 12]. Some of the major RFID applications are ID badges, supply chain management, inventory management, access tracking, and asset tracking. RFID is further classified into many categories: low frequency (LF-RFID), high frequency (HF-RFID), ultra high frequency (UHF-RFID), active RFID, passive RFID, and battery-assisted passive (BAP-RFID).

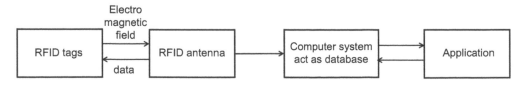

FIGURE 7.3
Working of RFID technology.

c. **Cloud Computing:** Eliminating the direct connection to a server, cloud computing delivers IT services by retrieving the resources from the Internet through web-based applications and tools. Cloud computing removes the need for physical resources, such as a hard drive; instead it makes it possible to store files in a remote intangible database [13]. A device with Internet access can have access to the files stored in the remote database as software is needed to run them. The information does not require a user to have access to a dedicated database or it to be in a specific place to gain access to the stored files [14]. The information is stored in the cloud, hence the name *cloud computing*. Clouds are classified into public cloud, private cloud, and hybrid cloud. Cloud computing services fall into four categories: infrastructure as a service (IaaS), platform as a service (PaaS), software as a service (SaaS), and serverless computing. Top benefits of cloud computing include cost, speed, performance, security, productivity, and elasticity [15]. Major service providers in this field are Google Cloud, Microsoft Azure, Amazon Web Services, Alibaba, and IBM Cloud.

d. **M2M: Machine-to-Machine Communication:** Being one of the foundations of IoT, M2M technology has found its application in many sectors, such as healthcare, infrastructure, business, agriculture, and many more. Its main purpose is to capture the sensor-tracked and collected data and transmit them to a network. It is similar to supervisory control and data acquisition (SCADA). The difference is that SCADA is a remote monitoring tool, whereas M2M technology uses public networks and access methods, such as Ethernet, making it cost effective [16]. M2M applications convert the data into automated actions. The main components of M2M systems are sensors, RFID, Internet communication links, and computing software. Telemetry is one of the earliest applications of M2M communication used for transmitting operational data. The many benefits of M2M communication include: (1) minimal equipment maintenance; (2) reduced costs, and hence increased profits in the business; and (3) improved customer services. Major M2M applications are shown in Figure 7.4.

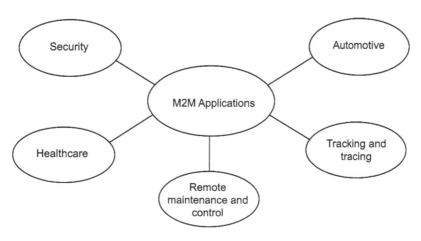

FIGURE 7.4
M2M applications.

7.4 Applications of IoT

IoT applications enhance the comfort of our lives by providing us with complete control over our routine, day-to-day tasks. The emerging technological breakthroughs provided by IoT offer promising solutions to the pressure created on healthcare systems, environmental requirements, and food supplies by the rapidly rising population all over the globe. IoT aims to bring the world together by providing a single application-specific platform to users. IoT finds its usage in almost all the applications used today globally. Some of the most widely used applications of IoT are discussed here [17, 18].

7.4.1 Smart Home

When we are not home, we are often crowded by many thoughts about what must be happening at our home in our absence. Did I switch off the fan? Did I turn the gas stove off? Did I switch off the water heater? Are the kids safe at home? Are they doing their homework? A smart home puts to rest all these worries and provides us with peace of mind. This can be achieved with just a quick glance at our smartphone or smart wearable [19].

IoT technology provides connection of the devices and appliances of our home with the cloud, i.e., the home network, so that they can communicate with each other and with us. Now the real question is, Which device of ours can behave in this manner? Any device at our home that functions using electricity can be positioned in the smart home network and expected to answer to our commands. These commands can be voice commands, remote commands, or smartphone or tablet commands. Currently, most smart home applications deal with home security, lighting systems, home theater and entertainment systems, and thermostat regulation [20]. Expectations for a smart home can be expressed as follows:

- Illuminate the path to the washroom during the night time.
- Monitor garbage in the trashcan and generate online orders for trashcan liner replacements.
- Suggest food recipes based on the ingredients stored in the refrigerator.
- Unlock the door automatically while approaching.
- Feed the pets at a regular interval based on a predefined schedule with a predefined amount of food content.
- Send texts or audio-message alerts when the laundry washing, spinning, or drying cycle ends.
- Instantly switch on the mood lighting for the current mood. Also instantly play songs based on the current mood.
- Preprogram the television so that it can be watched at certain times, especially for children.
- Temperature set (warm/cold based on current body temperature) the bedroom before waking up and also before coming out from a bath to match the body temperature. This is also useful when sick.
- Turn on/off the coffee maker from bed.

Smart homes are energy efficient as the devices show reduced levels of functionality when not in use. Based on the commands given by the user, the devices sleep and wake-up. Electricity bills also get lower as lights automatically turn off when not in use, and room temperature is also set based on whether someone is in the room at the moment. Smart devices are available to track how much energy is used by certain appliances and whether it is smart to continue using them. The prime beneficiaries of smart home technology are elderly people living alone. Smart home technologies can notify the elderly resident when to take medicines based on a predefined time schedule. Also, a smart home can alert the nearby connected healthcare center if the resident collapses or suffers from a heart attack (based on the symptoms or heartbeat captured by the wearable), can track how much and what the resident is eating, can track whether the resident is working out, and can remind the resident in case of forgetfulness to turn off the appliances and to lock the home while going out. The initial investment expenditure in a smart home is obviously large, but it is not as large as the overall expenditure carried out by the owner in the lifetime. It also allows children living away from home to take care of their aging parents. It also benefits differently abled people and people with limited range of movement [21].

7.4.2 Smart Cities

Smart city is one of most powerful applications of IoT attracting curiosity among the population all over the globe. It is a city equipped with basic infrastructure to give a decent quality of life and a clean, sustainable environment through application of some smart solutions [7]. Smart city has also become one of the major agendas of the political structure of a country. The Indian government aims to provide its citizens with at least one smart city in each state [7] and similar ideology is being carried out all around the world. Features of a smart city include robust Internet connectivity, e-governance, smart surveillance, smart energy management, smart water supply, automated transportation, smart security management, environmental monitoring, and many more [22].

The idea of smart city is tailored to solve the problems faced by the residents of major cities, such as traffic congestion, energy supply problems, pollution, garbage disposal, and so on. One example of smart city is smart trashcans enabled with cellular communication, which are capable of transmitting alerts to municipal service officials when a dustbin needs to be emptied. Another example can be informing the users who are driving cars about the available and free of cost (if needed) parking slots around the nearby area by installing sensors and making use of web applications.

7.4.3 Smart Retail

The retail sector holds enormous potential for IoT. It enables retail stores to transform their consumer-product-service relationships and provide consumers with a smooth shopping experience. It requires power-enabled, battery-free hardware and real-time digital analytics to manage smart retail functionalities, such as inventory management to manage out-of-stock products, and increasing operational efficiency. The most important challenge of smart retailing is to convert visitors into customers. In this tech-savvy generation, customers have become very smart as they compare the product before making up their mind and eventually end up buying that product online.

To address this problem, many retail chains are trying interactive media campaigns. The four pillars of interactive media campaigns are onboard, engage, convert, and retain. Here, the customer is categorized as a first timer, someone who comes occasionally, or someone who comes often. Also with respect to the time a customer spends in the store, it is categorized as a customer who spends less time in the store and a customer who spends a lot of time. Different messages are sent to different category customers. For example, if a customer visits for the first time, he/she can be given an extra discount to attract and then retain him/her: "Buy 2 shirts get 2 clothing items free." Suppose a customer spends a lot of time at a specific section of jewelry in the store; then a specific message can be sent to him/her: "Buy a pair of diamond earrings and get 30% discount on the making charges." This way of luring and retaining is customer-specific. The aim of smart retailing can be achieved by asset tracking, smart retail labels, and data analysis regarding the consumer-product-service relationship. Retailers connect with customers to improve and enrich the consumer-product relation experience. Store layouts, product positioning in high- and low-traffic areas in the store, product demand and supply ratio, and customer tracking are the main features of smart retaining. Retailers can be in constant connection with the consumers even when they are not in the store [23].

7.4.4 Smart Wearables

Smart wearables are hugely popular these days, and their market is booming. Hence, big companies like Samsung, Google, and Xiaomi are investing a lot of funds into the research and development (R&D), and marketing of these products. These devices currently are covering major markets of health, fitness, entertainment, and retail shopping requirements. Wearable devices are installed with sensors and software, which track and gather data and information about the users and their activities [24]. These collected data go through the complete cycle of processing to extract essential application-specific data about the user. For example, if the wearable is tracking the workout of the user, then it will collect all the data about the user's day-to-day activity (walking, running, swimming, climbing stairs, etc.) to predict and prescribe appropriate action to be taken by the user. In case of emergency during the workout, an alert message can be sent to the user (like reduce the treadmill speed due to increased heart rate), to the user's immediate family, and to a nearby healthcare center (in case of any health emergencies) so that appropriate action can be taken in time to avoid any loss.

The essential requirement from IoT technology of these smart devices is to be small-sized, energy-efficient, and low power consuming. A survey done by CCS predicts that the wearables market is bound to grow to $34 billion by 2020 [24]. Commonly wearable sensors are embedded in everyday products, such as watches, headbands, shoes, necklaces, and bracelets, making wearables attractive to the common public. Smart clothing is a novel field in which sensors are being attached and activity of the users is being tracked. These devices are becoming great gifting options these days.

7.4.5 Industrial Internet

Industrial Internet, also termed as Industrial Internet of Things (IIoT), is the new and hot topic of discussion in the industrial sector. The aim of IIoT is to empower the industrial sector with software, sensors, and big data analytics to create smart machines. IIoT is the current desirable investment in the industrial market and is an upcoming concept. The ideology behind IIoT is to create smart machines because they are more consistent, efficient, and accurate than humans while dealing with data, which in turn would help industries to identify problems and inefficiency much sooner to prevent any loss. IIoT holds great potential for quality control and sustainability [25], which in turn will increase the gross domestic product many times worldwide.

7.4.6 Connected Vehicles

There is a wave in the automobile industry to provide users with smart technology to enhance their in-vehicle experience by optimizing the internal functions of the vehicle. A connected vehicle, also termed a *smart vehicle*, is a vehicle that uses attached sensors and cloud computing and which is able to optimize its operations, maintenance, and comfort of passengers to provide them with a great overall experience. Major brand names, such as BMW, Tesla, Google, Apple, and some other automobile companies are working towards making smart vehicles a successful and affordable market [26].

7.4.7 Smart Grids

The concept of smart grid is becoming extremely popular all over the globe. The idea behind smart grids is to gather the data of electricity consumers is an automated manner to analyze the pattern of supply and consumption of electricity. This in turn will enhance the efficiency of electricity usage as well as help the economy in the long run [27].

7.4.8 Smart Farming: IoT in Agriculture

Being one of the most important fields of IoT, smart farming aims to meet the food-supply demands of the increasing population across the globe. Governments are investing funds into R&D of smart farming as well as bringing smart farming into practice, and helping farmers by providing them with advanced instruments and techniques to increase the crop production exponentially. Farmers also hope to yield better returns on investment [28]. Smart sensing will help farmers to analyze the moisture and nutrient contents of the soil, control water usage, and identify the better fertilizers from the options available to them.

7.4.9 Smart Healthcare: IoT in Healthcare

Smart healthcare is a heavily researched and sought-after field of IoT. It is already a reality in many parts of the world. Smart wearables are complementary to the field of smart healthcare. It not only contributes to the well-being of people by empowering them to lead a healthier life, but also provides companies with profit by bringing smart healthcare into the public domain. The data collected by smart wearables can be transferred to a smart healthcare system to help personalize an individual's health by prescribing custom-made schedules and strategies to combat health hazards. Immense R&D is being conducted in this field; many research papers, articles in journals, and conference proceedings are being published [29].

7.4.10 IoT in Poultry

IoT applications can be utilized for monitoring animal husbandry, to gather data about the health of the animals, and to send alert messages to ranchers and owners about sick animals. This field is still under research and yet to be implemented in many countries [30].

7.5 Conclusion

IoT applications are found in every field of life in today's world. We cannot imagine our lives without smart devices. This revolution has been brought about by IoT and its complements. It has made our lives pocket-friendly too. Cloud computing has dramatically changed the way we live our lives, adding comfort and convenience to many aspects. It has brought our world to function on a single platform. It has also provided various means of managing information. Its on-demand nature makes cloud computing available for use anywhere, anytime. Cloud computing, RFID, and M2M communication are the pillars that make IoT functional.

References

1. Gazis, V. (2017). A survey of standards for machine-to-machine and the Internet of Things. *IEEE Communications Surveys & Tutorials, 19*(1), 482–511.
2. Moeen, M., Page, A., Soyata, T., Sharma, G., Aktas, M., Mateos, G., Kantarci, B., & Andreescu, S. (2015). Health monitoring and management using Internet-of Things (IoT) sensing with cloud-based processing: Opportunities and challenges. In IEEE International Conference on Services Computing.
3. Egli, P. R. (2012). An introduction to MQTT, a protocol for M2M and IoT applications. Indigoo. com
4. Armbrust, M., Fox, A., Griffith, R., Joseph, A. D., Katz, R., Kowinski, A., … Zaharia, M. (2010). A view of cloud computing. *ACM Communication, 53*(4), 50–58.
5. Liu, Y., & Zhou, G. (2012). Key technologies and applications of Internet of things. In 2012 Fifth International Conference on Intelligent Computation Technology and Automation, pp. 197–206.
6. Botta, A. (2014). On the integration of cloud computing and Internet of things. In 2014 International Conference on Future Internet of Things and Cloud, pp. 23–30.
7. Shanbhag, R., & Shankarmani, R. (2015). Architecture for Internet of things to minimize human intervention. In 2015 International Conference on Advanced Computing Communication Informatics, pp. 2348–2353.
8. Babu, S. M., Lakshmi, A. J., & Rao, B. T. (2015). A study on cloud based Internet of Things: Cloud IoT. In 2015 Global Conference on Communication Technologies (GCCT), pp. 60–65.
9. Botta, A., Donato, W., Persico, V., & Pescape, A. (2016). Integration of cloud computing and Internet of Things: A survey. *Future Generation Computing System, 56*, 684–700.
10. Lee, I., & Lee, K. (2015). The Internet of things (IoT): Applications investments and challenges for enterprises. *BusinessHorizon, 58*(4), 431–440.
11. Singh., M., & Srivastava, V. M. (2018). Multiple regression based cloud adoption factors for online firms. In 2018 International Conference on Advances in Computing and Communication Engineering (ICACCE), Paris, pp. 147–152.
12. Alenezi, A., Zulkipli, N. H. N., Atlam, H. F., Walters, R. J., & Wills, G. B. (2017). The impact of cloud forensic readiness on security. In 7st International Conference on Cloud Computing and Services Science, pp. 1–8.
13. Zhou, J., Cau, Z., Dong, X., & Vasilakos, A. (2017). Security and privacy for cloud-based IoT: Challenges countermeasures and future directions. *IEEE Communication, 55*(1), 26–33.
14. Singh, M., Gupta., P. K., & Srivastav, V. M. (2017). Key challenges in implementing cloud computing in Indian healthcare industry. In 2017 Pattern Recognition Association of South Africa and Robotics and Mechatronics (PRASA-RobMech), Bloemfontein, pp. 162–167.

15. Atlam, H. F., Alenezi, A., Walters, R. J., & Wills, G. B. (2017). An overview of risk estimation techniques in risk-based access control for the Internet of things. In 2nd International Conference on Internet of Things Big Data and Security, pp. 1–8.
16. Díaz, M., Martin, C., & Rubio, B. (2016). State-of-the-art challenges and open issues in the integration of Internet of things and cloud computing. In *Journal of Network and Computer Applications*, pp. 99–117.
17. Dar, K. S., Taherkordi, A., & Eliassen, F. (2016). Enhancing dependability of cloud-based IoT services through virtualization. In 2016 IEEE First International Conference on Internet-of-Things Design and Implementation (IoTDI), pp. 106–116.
18. Atlam, H. F., Alenezi, A., Alharthi, A., Walters, R. J., & Wills, G. B. (2017). Integration of cloud computing with Internet of things: Challenges and open issues. In 2017 IEEE International Conference on Internet of Things (iThings) and IEEE Green Computing and Communications (GreenCom) and IEEE Cyber, Physical and Social Computing (CPSCom) and IEEE Smart Data (SmartData), Exeter, pp. 670–675.
19. Singh, M., et al. (2018). Cloud computing adoption challenges in the banking industry. In 2018 International Conference on Advances in Big Data, Computing and Data Communication Systems (icABCD), Durban, South Africa, pp. 1–5.
20. Bedi, G., Venayagamoorthy, G. K., Singh, R., Brooks, R. R., & Wang, K-C. (2018). Review of Internet of Things (IoT) in electric power and energy systems. *IEEE Internet of Things Journal*, 5(2), 847–870.
21. What is cloud computing: A beginner's guide. https://azure.microsoft.com/en-in/overview/what-is-cloud-computing/
22. Cloud Computing, Cybersecurity, Investopedia. (2019). https://www.investopedia.com/terms/c/cloud-computing.asp
23. Bedi, G., Venayagamoorthy, G. K., & Singh, R. (2016). Navigating the challenges of Internet of Things (IoT) for power and energy systems. In Clemson University Power System Conference (PSC), pp. 1–5.
24. Cruz, M. A. A., Rodrigues, J., Al-Muhtadi, J., Korotaev, V., & de Albuquerque, H. C. (2018). A reference model for Internet of things middleware. *IEEE Internet of Things Journal*, 5(2), 871–883.
25. Boyes, H., Hallaq, B., Cunningham, J., & Watson, T. (2018). The industrial internet of things (IIoT): An analysis framework. *Computer in Industry*, 101, 1–12.
26. Lu, N., Cheng, N., Zhang, N., Shen, X., & Mark, J. W. (2014).Connected vehicles: Solutions and challenges. *IEEE Internet of Things Journal*, 1(4), 289–299.
27. Miceli, R. (2013). Energy management and smart grids. *Energies, MDPI Journals*, 6, 2262–2290.
28. Muangprathub, J., Boonnam, N., Kajornkasirat, S., Lekbangpong, N., Wanichsombat, A., & Nillaor, P. (2019). IoT and agriculture data analysis for smart farm. *Computer and Electronic in Agriculture*, 156, 467–474.
29. Vippalapalli, V., & Ananthula, S. (2016). Internet of things (IoT) based smart health care system. In 2016 International Conference on Signal Processing, Communication, Power and Embedded System (SCOPES), Paralakhemundi, pp. 1229–1233.
30. Archana, M. P., Uma, S. K., & Raghavendra, B. (2018). Monitoring and controlling of poultry farm using IOT. *International Journal of Innovative Research in Computer and Communication Engineering*, 6(4).

8

Internet of Things and Artificial Intelligence

**Umang Kant[1], Mayank Singh[2], Shailendra Mishra[3],
and Viranjay M. Srivastava[2]**

[1]*Krishna Engineering College, Ghaziabad, Uttar Pradesh, India*
[2]*University of KwaZulu-Natal, Durban, South Africa*
[3]*Majmaah University, Majmaah, Kingdom of Saudi Arabia*

CONTENTS

8.1 IoT and AI—An Introduction

The upsurge in the field of computing can be seen as outdating of traditional technologies in the present and near future. IoT has made its presence felt in almost every area of computing, and in the IoT paradigm, the objects surrounding us fall into an arrangement of a network in some form. Sensor network technologies and radio frequency identification (RFID) meet this task by embedding information and communication systems within the environment surrounding us. The RFID group defines IoT as the universal network of interconnected objects uniquely addressable based on standard communication protocol [1]. This collaboration of IoT with various other components led to the generation of vast amounts of data, which are to be stored and processed in an efficient, user-friendly, presentable manner. A virtual infrastructure is needed for the management of such an environment, and cloud computing can be used for meeting this requirement.

Cloud computing can provide the virtual infrastructure for required computing that integrates monitoring data, devices, tools, platforms, and clients. Cloud computing offers end-to-end services used for managing businesses and users to access applications on demand at anytime from anywhere [1]. An essential part of IoT is the smart connectivity with available networks and application-based computing using the network and resources. Smart connectivity is being achieved by the progression of Wi-Fi and 4G/5G wireless Internet access. IoT aims at (1) attaining an in-depth understanding of user applications, (2) managing communication network models and application-based software architectures to process the user's requirements and information, and (3) achieving smart and autonomous behavior with the help of analytics tools of IoT.

As deliberated by the collection of European research projects on IoT [2]: "*Things* can be described as active participants in business, information, and social processes where they are made to interact and communicate among each other and with the environment by exchanging data and information sensed about the environment, while reacting autonomously to the real/physical world events and prompting it by running processes that trigger actions and create services with or without direct human intervention." A simple definition of a smart environment is given by Forrester [3], "a smart environment uses information and communication technologies to make the critical infrastructure components and services of a city's administration, education, healthcare, safety, transportation, and other utilities more cognizant, interactive, and efficient."

The IoT field requires the interdisciplinary nature of the technologies. As discussed by Atzori et al. [4], the usefulness of IoT can be implicit in the application domain at the intersection of the following three paradigms: (1) Internet oriented (middleware), (2) things oriented (sensors), and (3) semantic oriented (knowledge).

IoT requires sensors to be embedded into all kinds of devices, which provide streams of data through Internet connectivity to a central location. The sensors sense those data, which in turn are analyzed using analytics tools and further acted upon using other tools to ultimately provide services to users. Almost all IoT-related services follow five steps: (1) sense, (2) transmit, (3) store, (4) analyze, and (5) act [5]. An IoT application becomes worthy of investments of time and money only if it demonstrates the last step of these five steps: *act*. The application must act according to the data acquired and analyzed to provide complete service to the client. This act or action can range from merely providing customer-related information (e.g., sending an alert message of an accident to a family member of a user/or a nearby healthcare center) to a reflective physical action (e.g., deploying an ambulance to an accident site). This last step of the cycle is entirely dependent upon the analysis done by the analytics tools of the IoT-based software. Hence, it is at the fourth step, i.e., *analyze*, where the entire scope and value of the IoT-based service application is concluded. This is where AI or a subset of AI called *machine learning* (ML) plays a pivotal role.

A natural way to build an AI-enabled IoT system is to use a centralized approach [6], as depicted in Figure 8.1. The procedure involving model building in an AI environment involves a great amount of processing power as the most efficient models are developed with the help of large quantities of training data. The capability of handling huge amounts of data can be achieved using a central cloud repository or a large data center. Other than the one mentioned, centralization has other benefits too; a central system is always easier to manage and maintain. Also, a central system provides better security and resilience, and proves to be more economic. These and other supplement benefits of this centralized approach are responsible for the growth in the demand and popularity of IoT in the market.

There exists an indistinct overlap between IoT and AI. In this overlap, IoT connects machines, and captures/stores/analyzes the data acquired from these machines, whereas AI simulates the intelligent behavior in the machines connected in the IoT environment. To make logic out of the stored data generated by the devices, AI functionally deals with this huge amount of data to make the machine act in an intelligent manner. Data are only

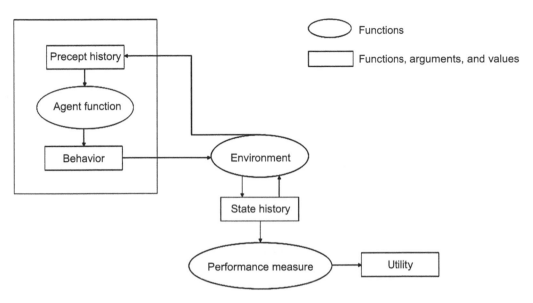

FIGURE 8.1
Basic concept of artificial intelligence.

useful if they create an action appropriate to the requirements by the user [7]. Hence, devices need to be connected in an intelligent manner rather than be just connected. AI caters to both real-time processing and post-event processing [7]. In real-time processing, the system needs to respond really quickly to situations (hard real-time and soft real-time), whereas in post-event processing the system needs to identify the patterns and clusters in the dataset by applying predictive analytics.

Machine learning (ML) provides the ability to perform pattern detection (process of detecting patterns) in the available datasets. It learns from the detected patterns, and adjusts the procedures that provide the medium for eliciting actions. ML supports the ability to automatically identify patterns and detect anomalies in the data acquired by the sensors. For example, decisions based on information, such as pressure, air quality, sound, traffic congestion, and so on, can be acquired using smart devices supporting ML and AI. Apart from ML, other AI technologies, such as computer vision and speech recognition, also help in making smart decisions in certain situations. IoT/AI-based applications support multinational companies to enhance operating efficiency, seed new product-based services, oversee risk management, evade unplanned losses, and fare the resources and in turn smart environments. IoT and AI aim to make the business world smarter by: (1) improving customer satisfaction, (2) managing risk management, (3) increment operational efficiency, and (4) enhancing the business analytics process [8]. Figure 8.1 shows the basic concept of AI.

8.2 IoT and Edge Computing

As discussed by Helder Antunes [9], senior director of corporate strategic innovation at Cisco, "in most scenarios, the presumption that everything will be in cloud with a strong and stable fat pipe between the cloud and the edge device is not realistic."

This statement indicates that we now need to incorporate *edge computing* as well for the efficient and productive business strategies catering to the needs of millions and millions of users connected over the Internet. We often discuss the explosion in the quantity of data being generated by smart devices. The traditional method of storing and processing all the data in the centralized cloud (as discussed above) has proved to be too costly and slow for meeting user requirements. These concern areas have encouraged researchers to move towards edge computing. Edge computing facilitates data processing closer to the source of the generated data. As suggested by Gartner Group [10], "Approximately 90% of the enterprise-generated data is processed at the traditional centralized data center or cloud, which means at present around 10% of this data is processed outside this traditional environment." Gartner predicts that, "by 2022, this figure of 10% will exponentially rise to 75%, i.e., about 75% of generated data will be processed outside traditional data centers or cloud and will have a profound impact on the enterprises."

As discussed above, edge computing permits data produced by IoT devices to be processed nearest to where it has been created instead of transferring the data to clouds (data centers) via long courses. Edge computing enables computing to be done closer to the edge of the networks, in turn authorizing organizations to do their analytics in real-time scenarios, e.g., finance, healthcare, manufacturing. According to research firm IDC [9], edge computing can be understood as a mesh network of micro data centers that store and process frequently used data or critical data locally. It drives all the received data to a central data hub, i.e., cloud storage repository. The edge devices collect a massive amount of data, which needs to managed and processed at the cloud (data center). Edge computing enables capturing some of the critical data to be processed locally, which in turn reduces the burden of data traffic at the central cloud repository. This capturing of the critical data is achieved by IoT devices transferring the data to those local devices that support network connectivity, storage, and processing in very small measures. The captured data are processed at the edge, and then either all the data or a small portion of it is sent to the central cloud repository or to a corporate data center.

Edge computing triumphs over cloud computing when real-time or time-sensitive events are to be executed, e.g., self-driving cars are required to react to external factors in real-time only [11], or railways or chemical plants need to meet real-time requirements. Edge computing complements cloud computing. Edge computing is ideal in many environments, but the situation in which it is the most sought-out is when it is not feasible for IoT devices to be continuously connected to the central cloud, i.e., when devices have poor connectivity issues. Another situation is when the environment has to deal with latency dependent processing. By latency dependent processing we mean, edge computing moderates latency by managing the data at a local device, i.e., the data do not travel to the cloud repository or corporate data center over the network for further processing.

When it comes to security, many researchers argue that security is better in edge computing as the data stay close to their source (where they are being created) and do not traverse over a network. And due to this reason, minimum data are vulnerable and at the risk of being compromised as there are fewer data to be managed in a cloud repository or data center in a given environment. On the other hand, another class of researchers argues that edge computing is less secure as the edge devices themselves are vulnerable. Hence, the design of any edge computing environment (edge devices) must be of supreme prominence.

8.3 IoT and Fog Computing

The term *fog* was originally coined by Cisco [12], but a more general definition is discussed by Yi et al. stating, "fog computing is a geographically distributed computing architecture with a resource pool which consists of one or more ubiquitously connected heterogeneous devices (including edge devices) at the edge of network and not exclusively seamlessly backed by cloud services, to collaboratively provide elastic computation, storage, and communication in isolated environments to a large scale of clients in proximity" [13]. The OpenFog Consortium [14] also gives a definition to fog computing as "a system-level horizontal architecture that distributes resources and services of computing, storage, control, and networking anywhere along the continuum from cloud to things." Similar to cloud and edge computing, fog computing too provides services such as data storage and data processing for IoT environments. Fog computing takes its inspiration from edge computing, i.e., instead of sending the data to the cloud, it provides data processing and local storage of the generated data to fog devices. This provides temporary storage and leads to reduced network traffic and latency. In contrast to the cloud, the fog provides services with faster response and greater quality [15].

The amalgamation of IoT with fog computing generates a new prospect for service, which can also be termed as fog as a service (FaaS). In FaaS, the service provider constructs a hub of fog nodes across the dedicated area, and the service provider then becomes the owner of the hub. In this hub, each fog node is responsible for local storage, computation, and networking capabilities enabling both small and big enterprises to deploy the computing to meet a wide variety of users' requirements in an efficient and fast manner. As depicted in Figure 8.2, fog computing provides an extension to cloud computing by acting as an intermediate between the cloud and the end devices; this results in bringing storage, processing, and networking services closer to the end devices (fog nodes) themselves. Fog computing acts as a bridge between the cloud and the end devices. A fog device must be capable of having network connectivity, processing, and storage (e.g., routers, embedded servers, smart cameras, controllers). Table 8.1 compares cloud computing and edge/fog computing.

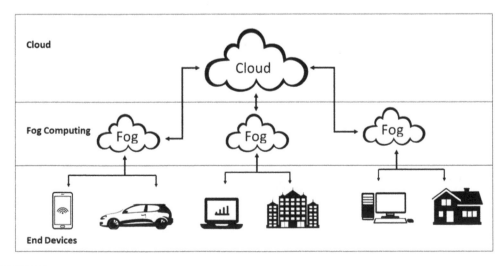

FIGURE 8.2
IoT with fog computing.

TABLE 8.1

A Comparison between Cloud Computing and Edge/Fog Computing

S. No.	Reference	Cloud Computing	Edge/Fog Computing
1	Latency	High	Low
2	Storage	Scalable	Limited
3	Processing power	Scalable	Limited
4	Server node location	Within the Internet	At the edge of local network
5	Client to server distance	Multiple hops	One hop
6	Security	Defined	Flexible
7	Deployment	Centralized	Distributed
8	Location awareness	No	Yes

After detailed discussion about IoT, cloud computing, edge computing, and fog computing, it may be concluded that all of them overlap and complement one another. One can be better than the other based on a particular application. Figure 8.3 shows the overlap of cloud computing, edge computing, and fog computing to IoT environments.

8.4 Applications

8.4.1 Healthcare and Activity Tracking

One of the key applications of IoT is in the field of healthcare. Hence, business has a keen eye on this sector in particular. Healthcare centers provide real-time computing and responses in critical situations. This has led to a large number of smart devices being launched into

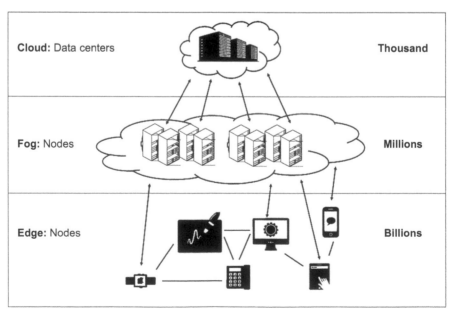

FIGURE 8.3

Overlap of cloud, edge and fog computing with IoT.

the market to cater to multiple requirements of all kinds of users. This application entails a reliable network connection in its environment in order to facilitate continuous processing, storage, and data retrieval from the cloud [16, 17].

8.4.2 Smart Traffic Lights

Fog computing and IoT enable traffic signals to manage the traffic and human movement based on the sensors attached to traffic lights. These sensors sense the flashing lights and make real-time decisions to manage vehicle and human movement. The sensors are capable of sensing the presence of pedestrians, two-wheeled, and four-wheeled vehicles, and measuring the distance and speed of pedestrians and vehicles. Smart traffic lights are considered to be fog nodes in the environment. A well-connected network is a paramount requirement for such applications [18, 19].

8.4.3 Smart Homes

These days markets provide users with many smart sensors and devices to be connected in the home. Though smart in nature, these devices support different platforms as they are independent and are from different service providers. Fog and edge computing integrates all the devices having different platforms and makes the smart home applications enabled with different resources. Fog computing provides a uniform interface for integrating different and independent devices, hence making the environment flexible [13, 19].

8.4.4 Connected Vehicles

IoT aims at making autonomous vehicles as a new generation of vehicles. The capabilities of smart vehicles include hands-free operations, automatic steering, and self-parking features reducing the human intervention to be bare minimum. It also includes the ability to communicate with nearby vehicles and the Internet network of the environment. This application requires soft real-time interactions. Another goal is to avoid collisions and accidents, which can be achieved through fog computing as it reduces latency [19, 20].

8.4.5 Augmented Reality

Augmented reality (AR) refers to those systems that add visual information to real-world scenarios. AR applications are latency sensitive as computing devices have become faster and smaller; hence, even a minor delay in the response leads to a catastrophic result. Fog computing employs fog and cloud computing to enable real-time responses in various situations [21–23].

8.5 Edge/Fog Deployment on IoT Platforms

Although beneficial, fog computing faces many challenges with its deployment. For successful integrating of fog with IoT, all the problems must be sorted out in advance. These challenges are discussed in brief in Table 8.2.

TABLE 8.2

Challenges Being Faced While Implementing Fog and IoT

S. No.	Challenge	Description
1	Scalability [24, 25]	IoT devices number in the millions and trillions, generating huge amounts of data. This large amount of data also requires huge amounts of storage and processing power. Edge/fog servers should be capable of supporting, managing, and responding to these devices along with their requirements in short time frames.
2	Latency [25]	High latency was the major reason for the shift from the cloud paradigm to the fog paradigm. The fog paradigm must be able to provide low latency especially for time-sensitive applications.
3	Complexity [24]	Edge/fog computing environments must be capable of dealing with the complexities posed by a large number of different devices. These devices and dedicated sensors are designed and manufactured by different manufacturers making the configuration and deployment process very complex. All the protocols must be properly followed while configured.
4	Energy consumption [26]	One major challenge to be addressed is maintaining energy consumption. Edge/fog environments involve a large number of devices (as discussed), and because their computation is distributed, unlike centralized in the cloud model, it proves to be less energy efficient.
5	Security [13, 19, 24, 27]	As compared to the cloud model, edge/fog computing is more vulnerable and less secure. The existing privacy and security measures of the cloud model cannot be directly applied to the edge/fog model as the fogs are deployed in a distributed manner and are not centralized. Researchers suggest authentication and cryptography to achieve the required network security. One plus point is that all the fog nodes must be protected using the same protocols, procedures, and controls.
6	Dynamicity [24]	IoT devices dynamically alter their workflow as they evolve with time. Though dynamic (an advantage), it results into degradation in performances and alternations in the internal properties of the devices. Another issue is that these devices suffer from hardware and software aging resulting in shifts in behavior and the workflow. Hence, the fog model is expected to be capable of automatic and intelligent reconfiguration of devices and the structure of the environment.
7	Resource management [17, 28]	Large numbers of devices and their requirements are met using many resources in the fog computing paradigm. Smart management of fog resources is expected for efficient functioning of the fog computing paradigm.

8.6 Conclusion

As evident, IoT has exponentially expanded its reach in every sector of the market. It has become an integral part of everyone's life as it connects everyone to everything in the given environment. Although IoT devices have low storage and processing capabilities, they still are better as compared to traditional centralized cloud architectures, which suffer from many issues such as high latency and network crashes. To overcome these issues, edge and fog computing are being adopted. Edge and fog computing complement the cloud and are in closer proximity closer to the end devices. As a result, edge and fog computing perform the processing near the end devices and reduce latency and network crashes, especially in time-critical real-time applications. The integration of IoT with AI and edge and fog computing is seen as a benefit to IoT applications at a larger scale. This chapter mainly focuses on delivering a brief introduction on IoT with AI, edge computing, and fog computing, and the comparisons between cloud computing and edge and fog computing. The discussions also present five major applications making use of either edge or fog computing in IoT environments. At the end, the chapter touches on the challenges in deploying edge or fog computing in IoT-based applications.

References

1. Gubbi, J., Buyya, R., Marusic, S., and Palaniswami, M. (2013). Internet of Things (IoT): A vision, architectural elements, and future directions. *Future Generation Computer Systems*, 29(7), 1645–1660.
2. Sundmaeker, H., Guillemin, P., Friess, P., & Woelfflé, S. (2010). Vision and challenges for realizing the Internet of Things, Cluster of European Research Projects on the Internet of Things. In CERP IoT, Cluster of European Research Projects on the Internet of Things.
3. Belissent, J. (2010). Getting clever about smart cities: New opportunities require new business models. *Forrester Research*.
4. Atzori, L., Iera, A., & Morabito, G. (2010). The Internet of Things: A survey. *Computer Networks*, 54(15), 2787–2805.
5. Brendan, O. (2015). Why the IoT needs artificial intelligence to succeed. In insidebigdata.com. https://www.ariasystems.com/blog/iot-needs-artificial-intelligence-succeed/
6. Calo, S. B., Touma, M., Verma, D. C., & Cullen, A. (2017). Edge computing architecture for applying AI to IoT. In 2017 IEEE International Conference on Big Data (Big Data), Boston, MA, pp. 3012–3016.
7. Vashi, S., Ram, J., Modi, J., & Verma, S. (2017). Internet of Things (IoT): A vision, architectural elements, and security issues. In 2017 International Conference on I-SMAC (IoT in Social, Mobile, Analytics and Cloud) (I-SMAC), Palladam, pp. 492–496.
8. Jones, M. (2018). How IoT and Artificial Intelligence will Make your Business Smarter. *Analytics Insights*.
9. Butler, B. (2017). What is edge computing and how it's changing the network. *Network World*.
10. Aktaş, I. (2018). Why edge computing for IoT? *Bosch Connected World Blog*. https://blog.bosch-si.com/bosch-iot-suite/why-edge-computing-for-iot/
11. Simone, E. (2018). Why do we need edge computing? *IoT for all*. https://www.iotforall.com/edge-computing-benefits/
12. Bonomi, F., Milito, R., Zhu, J., & Addepalli, S. (2012). Fog computing and its role in the internet of things. In Proceedings of the first edition of the MCC workshop on Mobile cloud computing, Finland, pp. 13–16.
13. Yi, S., Hao, Z., Qin, Z., & Li, Q. (2015). Fog computing: Platform and applications. In 2015 Third IEEE Workshop on Hot Topics in Web Systems and Technologies (HotWeb), Washington, DC, pp. 73–78.
14. Al-Doghman, F., Chaczko, Z., Rakhi, A., & Klempous, R. (2016). A review on fog computing technology. In 2016 IEEE International Conference on Systems, Man, and Cybernetics (SMC), Budapest, pp. 1525–1530.

15. Atlam, H. F., Walters, R. J., & Wills, G. B. (2018). Fog computing and the Internet of things: A review. *Big Data and Cognitive Computing, 2*(10), 1–18.
16. Nikoloudakis, Y., Markakis, E., Mastorakis, G., Pallis, E., & Skianis, C. (2017). An NF V-powered emergency system for smart enhanced living environments. In 2017 IEEE Conference on Network Function Virtualization and Software Defined Networks (NFV-SDN), Berlin, pp. 258–263.
17. Dastjerdi, A. V., & Buyya, R. (2016). Fog computing: Helping the Internet of Things realize its potential. *IEEE Computer, 49*(8), 112–116.
18. Peter, N. FOG computing and its real time applications. *International Journal of Emerging Technology and Advanced Engineering, 5*(6), 266–269.
19. *Fog Computing and the Internet of Things: Extend the Cloud to Where the Things Are* (White Paper). (2016). http://www.cisco.com/c/dam/en_us/solutions/trends/iot/docs/computing-overview.pdf
20. Adhatarao, S. S., & Arumaithurai, M. (2017). FOGG: A fog computing based gateway to integrate sensor networks to Internet. In 2017 29th International Teletraffic Congress (ITC 29), Genoa, pp. 42–47.
21. Dastjerdi, A. V., Gupta, H., Calheiros, R. N., Ghosh, S. K., & Buyya, R. (2016). Fog computing: Principles, architectures, and applications. In *Internet of things: Principles and paradigms* (pp. 61–75). Morgan Kaufmann Publishers Inc.
22. Kim, S. J. J. (2012). A user study trends in augmented reality and virtual reality research: A qualitative study with the past three years of the ISMAR and IEEE VR Conference Papers. In 2012 International Symposium on Ubiquitous Virtual Reality, Adaejeon, pp. 1–5.
23. Zao, J. K., Gan, T. T., You, C. K., Mendez, S. J. R., et al. (2014). Augmented brain computer interaction based on fog computing and linked data. In 2014 International Conference on Intelligent Environments, Shanghai, pp. 374–377.
24. Luan, T. H., Gao, L., Li, Z., Xiang, Y., Wei, G., & Sun, L. (2015). Fog computing: Focusing on mobile users at the edge. In arXiv.
25. Choi, N., Kim, D., Lee, S-J., & Yi, Y. (2017). Fog operating system for user-oriented IoT services: Challenges and research directions. *IEEE Communications Magazine, 55*(8), 44–51.
26. Ni, J., Zhang, K., Lin, X., & Shen, X. S. (2018). Securing fog computing for Internet of things applications: Challenges and solutions. *IEEE Communications Surveys & Tutorials, 20*(1), 601–628.
27. Mukherjee, M., Matam, R., Shu, L., Maglaras, L., Ferrag, M. A., Choudhury, N., & Kumar, V. (2017). Security and privacy in fog computing: Challenges. *IEEE Access, 5*, 19293–19304.
28. Mouradian, C., Naboulsi, D., Yangui, S., Glitho, R. H., Morrow, M. J., & Polakos, P. A. (2018). A comprehensive survey on fog computing: State-of-the-art and research challenges. *IEEE Communications Surveys & Tutorials, 20*(1), 416–464.

Part 3

IoT with Visual Surveillance for Real-Time Applications

9

Internet of Things-Based Speed and Direction Control of Four-Quadrant DC Motor

Bhupesh Kumar Singh and Vijay Kumar Tayal
Amity University, Noida, Uttar Pradesh, India

CONTENTS

9.1 Introduction

For general purposes, a substantial number of motors are being used, from those in our neighborhoods and households to machine tools in industries. The electric motor is now a fundamental and important source of power in the industries. These motors are required to perform a wide range of functions. Direct current (DC) motors are commonly used to convert DC (electrical energy) into mechanical energy with superior speed-control characteristics. The basic principle is based on Fleming's left-hand rule. The main purpose of this system is to develop a model for wireless control operations of a four-quadrant DC motor; these operations include speed and direction. Unlike conventional two-quadrant DC motors, special-purpose four-quadrant DC motors [1–3] are used to provide a wide range of speed control, quick braking, and operation in both forward and reverse directions. These motors are used in industries to lift heavy loads or to drive carriage trolleys. The advantages of DC motors over their alternating current (AC) counterparts are higher starting torque with simpler construction. The only major problem with the operation of DC motors is that their commutation process leads to limited usage of these motors in industry. However, DC motors run better than any other machines for four-quadrant operations. A four-quadrant motor operates as follows:

a. **Forward drive:** In quadrant 1, voltage and current are positive, and both rotation and torque are clockwise. Therefore, the motor runs forward [4].

b. **Reverse drive:** Here voltage and current are negative, so rotation is counterclockwise and torque is counterclockwise. So, the motor runs backwards [4].

c. **Forward braking:** Here current is negative and voltage is positive. This means that the rotation is happening clockwise but now the torque is applied counterclockwise. So, the motor stops eventually [4].

d. **Reverse braking:** Here current is positive and voltage is negative. This means the rotation is happening counterclockwise but now the torque is applied clockwise. So the motor stops eventually but in a reverse direction [4].

The Internet of Things (IoT) is an arrangement of interrelated devices requiring interaction between human–human or human–computer. It covers an enormous scope in domestic and industrial control and communication applications. To communicate and control, the IoT is primarily based on cellular 4G, global system for mobile communications (GSM), Wi-Fi, Bluetooth, Zigbee, ANT, and radio frequency identification (RFID) systems [4–5]. In this work, a radio frequency (RF) transmitter and receiver module is preferred, as this system is cost-effective and has low power consumption. The transmitter consumes no battery power when transmitting logic zero. In the proposed model, pulse width modulation (PWM) technique and H-bridge are employed for wireless speed and direction control, respectively. The combination of PWM and H-bridge results in the most efficient, reliable, and economical operation. The two sections are the transmitter and receiver. The RF transmitter used here acts as a remote control. It can be used for adequate range (up to 200 meters). The receiver section decodes the commands sent by the transmitter before feeding the commands to the microcontroller (MCU), which has latched outputs to drive a DC motor via motor driver IC for required tasks. This chapter is organized as follows: After Section 9.1's introduction, Section 9.2 describes speed and direction control. Hardware design and block diagrams are mentioned in Section 9.3. Regenerative braking is covered in Section 9.4, practical implementation is described in Section 9.5, and the conclusion is presented in Section 9.6.

9.2 Speed and Direction Control

The proposed IoT-enabled wireless control scheme uses the PWM technique for controlling the speed of the motor [6–8].

9.2.1 Pulse Width Modulation

There are various other methods, but PWM is used here because it decreases the energy consumption and makes the operation energy-efficient because it does not generate any additional heat. It also improves the reliability of the motor because the motor does not run at full speed all the time [9–11].

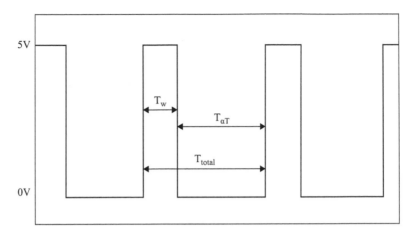

FIGURE 9.1
Pulse width modulation (amplitude vs time) representation.

PWM works on the binary method, i.e., it has only two signal periods (high and low). The main concept used in PWM is to vary the duty cycle (ratio of T_w to T_{total}) as shown in Figure 9.1. Either of the two can be varied to control the speed. In this case, T_w is being varied.

9.2.2 H-Bridge

In general, an H-bridge is a simple circuit consisting of four switching elements, i.e., transistors with the load at the center. The whole configuration depicts an H-like structure. The switching elements (Q1–Q4) are generally bipolar transistors, and diodes (D1–D4) are the Schottky type. The operation of an H-bridge is fairly simple. If Q1–Q4 are being turned on, the motor is going to move forward, and if Q2–Q3 are being turned on, the motor is going to reverse its direction. This is how the direction in the proposed model is controlled. However, to avoid a short circuit, care should be taken so that Q1–Q2 or Q3–Q4 are not closed simultaneously.

The H-bridge control sequences for various motor operations such as short circuit, clockwise motoring, counterclockwise motoring, and braking are shown in Table 9.1.

TABLE 9.1

H-Bridge Control Sequence

Q1	Q2	Q3	Q4	Operation
1	1	1	1	Short-circuit
1	0	0	1	Clockwise motoring
0	0	0	0	Coasting
0	1	1	0	Counterclockwise motoring
1	0	1	0	Braking
1	1	0	0	Short-circuit
0	1	0	1	Braking
0	0	1	1	Short-circuit

9.3 Hardware Design and Block Diagram

To simplify the IoT-based wireless speed control circuit design, the hardware is divided into three main sections: transmitter section, receiver section, and the receiver signal accepting circuit. Figure 9.2 and Figure 9.3 show the overall system design for the four-quadrant speed control of DC.

9.3.1 Transmitter Section

The transmitter section includes an RF-module transmitter that receives signals from the encoder and sends them to the receiver section for speed control (see Figure 9.2).

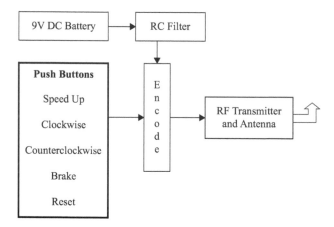

FIGURE 9.2
Transmitter signal generation circuit.

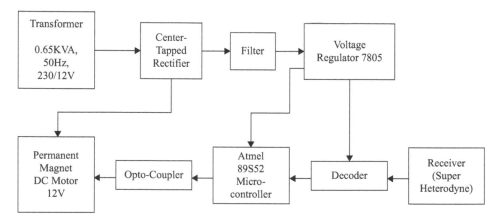

FIGURE 9.3
Receiver signal accepting circuit.

In the transmitter section, a 6-volt supply is passed through a rectifier (DC purpose) to an MCU.

A switch control is used to give inputs to the MCU for speed control. The MCU works on a specified coding and gives the output that is fed to the encoder. The encoder converts the input to desired output, and the output of the encoder is sent to the RF module. The RF module sends a signal of radio frequency to the receiver section.

9.3.2 Receiver Section

In the proposed model, two MCUs are used to control the speed of the motor.

Primarily, the power supply of the receiver section (Figure 9.3) consists of a combination of a transformer (step-down), rectifier (AC-DC), and the regulator (IC 7805) to finally give the output magnitude of 5 volts. The receiver gets the signal from the transmitter and sends it to the decoder (HT12 decoder). The signal is decoded in the decoder and is sent to the MCU (AT89S52) in the desired form. The MCU operates on the input and sends digital output signals to the motor driver (L293D) that is a dual H-bridge connected with a DC motor. The signals sent to the motor from the motor driver are used in final speed and directional control operations.

9.4 Regenerative Braking

Regenerative braking occurs whenever the speed of the motor is more than that of the synchronous speed. In this method, the motor runs as a generator, and the load is used to provide the required power to the supply [12–14]. The main criterion for regenerative braking is that the speed of the motor becomes greater than the synchronous speed. This condition will lead to the motor acting as a generator and the direction of both the current and torque reversing. This method can be utilized where the load on the motor has very high inertia.

When applied voltage to the motor is less than the back electromotive force (EMF), then both the armature current as well as armature torque reverses and the speed reduces. As generated EMF exceeds the applied voltage, the power transport takes place from the load to the supply.

9.5 Practical Implementation

The final implementation of the four-quadrant speed and direction control of the DC motor [15–18] is shown in Figure 9.4 and Figure 9.5. The receiver and transmitter sections are shown in Figure 9.4 and Figure 9.5, respectively. The hardware is designed and the testing is completed using the MCU programming, and the results are verified as per the four-quadrant operation of the DC motor. The various components used, along with their ratings are described in Table 9.2.

FIGURE 9.4
Receiver section.

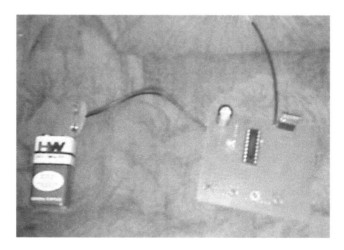

FIGURE 9.5
Transmitter circuit with AT89S52.

TABLE 9.2

DC Motor Components Used with Ratings

Component Name	Component Rating
Transformer	230/12 V, 0.65 KVA
RF module	5V, 437 MHz, 200 Mts
Center-tapped rectifier	12 V–12 V
Voltage regulator	7805
Encoder/decoder	5 V each
DC voltage source	9 V (receiver panel)
RC filter circuit	50 Hz
H-bridge	5 V
Opto-coupler	—

9.6 Conclusion

This chapter described IoT-based speed and direction control of four-quadrant DC motors using RF mode. The aim is hardware design of a wireless DC motor speed and direction control system. The PWM is used in conjunction with H-bridge as it is more effective, reliable, efficient, and economical than any other strategies. MCU AT89S52 is used to establish an interface between the transmitter and the receiver section. The model is best suited for small motors due to their low inertia. However, with suitable changes it can be used for large motors at the industrial scale. The H-bridge may be replaced by DC choppers if required for larger size motors. The hardware is designed and tested, and the real-time results are verified. The program is highly efficient, and the results obtained are encouraging. The designed power and control circuit works appropriately and fulfills the application requirements. The above model, if realized on a large scale, can enhance vehicle safety by reducing the stopping distance and immediately reversing the direction of the motion. For regenerative braking to be applied, the load must have a high inertia that can be converted to produce enough electrical energy. The energy accumulated through regenerative braking can then be utilized to supply electricity to the computer consoles and to all the safety sensors installed in the vehicle.

Acknowledgment

The authors would like to thank the Department of Electrical and Electronics Engineering, Amity School of Engineering and Technology (ASET), Amity University Uttar Pradesh, Noida, India, for providing the laboratory facilities for testing and related works.

References

1. Bimbhra, P. S. (2015). *Electric machines I*. New Delhi: Khanna Publishers.
2. Kothari, D. P., & Nagrath, I. J. (2013). *Electric machines*. New Delhi: Tata McGraw-Hill Education.
3. Wadhwa, C. L. (2013). *Power systems*. New Delhi: New Age International.
4. Postscapes. (2019). *IoT Technology Guidebook*. https://www.postscapes.com/internet-of-things-technologies/
5. Elprocus (2019). *RF Module—Transmitter and Receiver*. https://www.elprocus.com/rf-module-transmitter-receiver/
6. Herman, S. (2010). Industrial motor control (6th ed.). Delmar, Cengage Learning.
7. Jiancheng, F., Xinxiu, Z., & Gang, L. (2012). Instantaneous torque control of small inductance brushless DC motor. *IEEE Transactions on Power Electronics*, 27(12), 246–258.
8. Pillay, P., & Krishnan, R. (1989). Modeling, simulation, and analysis of permanent-magnet motor drives, part II the brushless DC motor drive. *IEEE Transactions on Industry Applications*, 25(2), 274–279.
9. Dubey, G. K. (2001). *Fundamentals of electrical drives* (2nd ed.) (pp. 271–277). New Delhi: Narosa Publishing House.

10. Bimal, K. B. (2002). *Modern power electronics and AC drives* (pp. 483–495). New Delhi: Pearson Education Publications.
11. Krishnan, R. (2002). *Motor drives modeling, analysis and control* (1st ed.) (pp. 513–615). Prentice Hall of India.
12. Hua, C. C., & Kao, S. J. (2011). Design and implementation of a regenerative braking system for electric bicycles based on DSP. In Proceedings on 6th IEEE Conference on Industrial Electronics and Applications, pp. 703–707.
13. Railway Gazette International. (2007, July 2). Regenerative braking boosts green credentials. https://www.railwaygazette.com/news/policy/single-view/view/regenerative-braking-boosts-green-credentials.html
14. Raworth, A. (2014). Regenerative control of electric tramcars and locomotives. *Proceedings of the Institution of Electrical Engineers 1906–1907, 38,* 374–398.
15. Ohio Electric Motors. (2011). DC Series Motors: High Starting Torque but No-Load Operation III.
16. Laughton, M.A., & Warne, D. F. (2003). Electrical engineer's reference book (16th ed.). Newnes.
17. Yeadon, W. H., & Yeadon, A. W. (2001). *Handbook of small electric motors.* McGraw-Hill Professional.
18. Leno, J. (2017, May 1). The 100-year-old electric car. *Popular Mechanics.*

10

Design of Arduino-Based Smart Digital Liquefied Petroleum Gas Detector and Level Indicator Using Internet of Things Technology

S. Ranjan, Vijay Kumar Tayal, and Saket Kumar

Amity University, Noida, Uttar Pradesh, India

CONTENTS

10.1 Introduction

Liquefied petroleum gas (LPG) is widely used in many domestic applications, industries, and vehicles as a source of fuel. LPG has many desired characteristics, such as high calorific value, lesser soot, and less smoke produced. Thus, it does not provide as much harm to environment. Leakage of LPG has always been a major concern; therefore, protection systems must be implemented to safeguard against LPG in households, places of education, and work. Propane and butane are highly flammable chemicals present in LPG. Due to LPG's odorless nature, ethane oil is added to serve as powerful odorant. This helps in leakage detection.

LPG, also known as auto gas, is used for many purposes, such as heating and cooking [1–3]. It can also be used as an alternate source of fuel. For residential places, commercial premises, and vehicles running on LPG, gas leakage is a major concern. Sometimes people do not detect the leakage of LPG due to its faint smell, which is difficult to recognize instantly. To overcome these associated dangers, a gas leakage sensor with a very high sensitivity to propane and butane [4–6] can be installed at the locations where chances of gas leakage/fire are high. In the event of LPG leakage, gas sensors detect the leakage, send an SMS to alert the consumer, and switch on the LED and buzzer alerts.

Industrial applications have the highest chances of gas leakage, and this may be a major source of health hazards for employees. To maintain a safe environment at such places, continuous monitoring using IoT-controlled relays and actuators is required. A gas leakage sensor can be installed to detect gas leakage at various locations, such as in service stations, storage tanks, and vehicles running on LPG [7]. The minute gas leakage is detected by a sensor, or the low signal is identified by the microcontroller (MCU), alarm devices are activated by the MCU. After a few seconds, the exhaust fan is also turned on automatically.

In industrial applications, wireless sensor networks have been explored for LPG leakage detection and monitoring as a cheaper and more efficient option [8–11]. Low-cost wireless sensors and an Arduino controller in combination with IoT-enabled computers can be used to monitor and control various conditions in homes using ZigBee and Visual Basic. The real-time demand is displayed on a personal computer [12–17]. In many countries, LPG production is not adequate, and hence may not be supplied to users through pipelines. To ensure the safety of customers, an intelligent system, known as safe from fire (SFF), has been assembled with multiple actuators and sensors [17–19]. The system is controlled by an MCU. Multiple sensors provide the signal to the MCU and integrated fuzzy logic system. In addition to activating the home fire alarms, the SFF system informs the fire service department automatically if any fire hazard takes place.

At present, LPG customers have no way to determine the exact status of LPG in a cylinder. This leads to frequent delays in gas booking. Via an interactive voice response (IVR) system, users are asked to respond to certain prerecorded questions, which might confuse persons who are illiterate; as a result, gas booking becomes a tedious task for them. Further, the dedicated phone numbers provided by gas suppliers are busy due to a high volume of calls. The existing systems of leakage detection have major drawbacks, such as incapability to react on time, unavailability of data in case of an accident, and inaccurate determination of locations.

The IoT is a system of interrelated devices requiring human–computer or human–human interaction [20–21]. The IoT is commonly used to monitor and control the mechanical, electronic, and electrical systems used in private and public buildings, institutions, and industrial applications. The majority of the IoT relies on global system for mobile communications (GSM), Bluetooth, and Wi-Fi connectivity. Selecting which connectivity to use is completely based on the applications of the IoT product being created and user needs.

In this work, GSM-based connectivity is chosen, as this system is widely adopted due to its low hardware costs and good roaming capabilities. The GSM is the mobile telephone system meant for voice, but also supports data in the form of SMS and GPRS. The GSM is a wireless technology and has the possibility to connect remote areas. This technology is highly efficient, easy to handle, and user-friendly. An LPG weight and leakage detection system using GSM may be useful in various applications in homes and hotels.

The main aim of this work is to design a cost-effective alarm system that continuously functions to automatically detect and prevent the leakage of LPG. The proposed alarm system is very sensitive to butane. It consists of protection circuitry with an exhaust fan and an LPG-safe solenoid valve. In this, the gas booking system is automated with no human involvement required. The weight of the LPG cylinder is continuously measured; if the weight goes below a certain level, automatically a message is sent to the LPG supplier for timely delivery of a new gas cylinder. The GSM module alerts the users by sending them a warning message. Along with automatic delivery, security features are also incorporated. To avoid any major accident in the event of LPG leakage, the system monitors in real-time and provides a warning signal to the customers. The proposed technique may also detect leakage of toxic gases. The circuit will show the concentration of gas in the air and activate a warning system if the concentration exceeds safe limits. The major advantages of this system are its quick detection time and accuracy.

10.2 Hardware and Circuit Description

The major components used in this work include: (1) Arduino mega, (2) load cell, (3) LPG sensor, (4) GSM module, (5) keypad module, (6) LCD, (7) buzzer, (8) switch, and (9) adaptor. The proposed system consists of: (1) GSM module, (2) level-sensing module, (3) leakage module, and (4) display module.

10.2.1 GSM Module

The GSM module (Figure 10.1) is used for sending SMS to the consumer in the event of low LPG levels or any leakage in the cylinder. The GSM module has L-shaped contacts on four sides that need to be soldered on the side as well as on bottom. A SIM card holder is attached to the outer wall of the GSM module. Once a leakage is detected, an alarm is raised, and a buzzer is activated, while at the same time an SMS is sent to the mobile numbers stored in the system. When the LPG level reaches below a minimum level, an SMS is sent automatically to the distributor to book another cylinder.

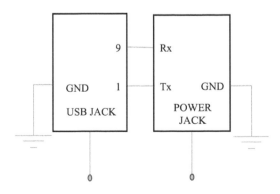

FIGURE 10.1
Interfacing of GSM module with Arduino.

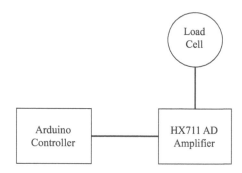

FIGURE 10.2
Load cell connected with HX711 amplifier.

10.2.2 Level-Sensing Module

A HX711 voltage amplifier is used to extract measurable data from a load cell designed for weigh scales. In this module, four load cells are connected to an HX711 voltage amplifier, and the HX711 voltage amplifier (Figure 10.2) is then connected with Arduino. This module effectively measures the content of gas in an LPG cylinder while displaying it on the LCD.

10.2.3 Leakage Detection Module

The leakage detection module mainly consists of an MQ6 sensor that is used to detect gas leakage. The MQ-6 gas sensor utilizes a tin dioxide-sensitive material with low conductivity in fresh air. If any combustible gas is produced, the conductivity of the sensor increases. The MQ-6 sensor can very quickly detect an LPG leakage, raise an alarm, and activate the connected buzzer. The interfacing of an MQ-6 sensor with Arduino is displayed in Figure 10.3.

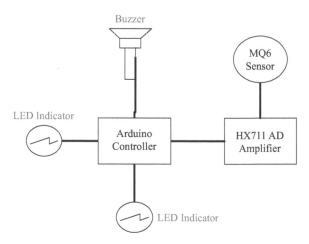

FIGURE 10.3
Interfacing of MQ6 sensor with Arduino.

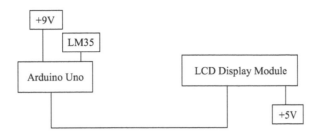

FIGURE 10.4
Interfacing of display module with Arduino.

10.2.4 Display Module

The interfacing of an LCD module with Arduino is very important in designing such embedded systems (Figure 10.4). The JHD162A module is used as an LCD for representing the level and leakage of gas in the cylinder. The LCD displays the concentration of gas and leakage in the cylinder (Figure 10.5).

10.3 Algorithm of Proposed Technique

The Arduino programming algorithm is very simple. It involves the interfacing of the sensor, buzzer, GSM, and GLCD.

10.3.1 Interfacing of MQ6 Sensor

The output of the gas sensor is displayed on the serial monitor. The value of the gas sensor varies between 0 to 1024. This value is an analog variable thus an analog pin of Arduino is used. To activate the serial monitor, a baud rate of 9600 is set. The Arduino reads the output of the sensor using function *analog read* and displays it on the serial monitor.

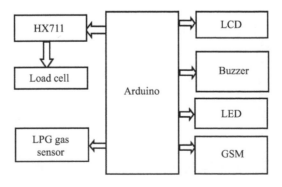

FIGURE 10.5
LPG detector and level indicator schematic diagram.

10.3.2 Interfacing of Buzzer

The buzzer's positive wire is connected to digital pin number 3. Digital pins always give logical 1 or 0. This means a pin will either remain in a state of high or low. From the above interfacing, the buzzer will buzz at an interval of 100 milliseconds. Using *pinMode()* function, the buzzer pin is set as an output pin.

10.3.3 Interfacing of GSM

GSM interfacing is basically a serial port communication with Arduino. The library file *SoftwareSerial.h* for GSM interfacing is required. GSM interfacing is always done with serial ports. One can set the pins of Arduino as serial ports through the function *mySerial()*, and the communication between Arduino and GSM begins. The ATtention (AT) modem commands are used to initiate the GSM in text mode. The AT+CMGF (command name in text: message format) command sets GSM module in messaging mode to send the message. The program allows users to send and receive SMS on the serial monitor.

10.3.4 Interfacing of GLCD

In GLCD, patterns are drawn using various syntaxes. This indicates welcome screen, level indicator, and percentage indicator.

10.4 Results and Discussions

10.4.1 Welcome Screen

Smart LPG Trolley is a welcome screen that appears on power activation (Figure 10.6). When power is connected through the system, it will show the name of the prototype. After the name of the project appears in the GLCD, the menu and functionality of the work are displayed in the GLCD. The first screen of functional work is shown in Figure 10.7.

The screen has four modes of operation. By setting these modes, users can easily access and adjust the parameters to work. Different modes of operation shown on the screen are explained next.

FIGURE 10.6
Welcome screen on GLCD.

FIGURE 10.7
First functional screen of Smart LPG Trolley.

10.4.2 Mode A

Mode A activates the function to enter the empty cylinder weight and net gas weight (Figure 10.8) written on the empty cylinder. Using the keypad, users can enter the respective weight of the cylinder.

As soon as the weight of the empty cylinder is entered, the user needs to press the D button to get access to enter the weight of gas in the cylinder (Figure 10.9). Thereafter, D must be pressed again to make an exit from mode A.

10.4.3 Mode B

Once the user exits mode A, button B needs to be pressed to get access to mode B. Mode B allows users to enter their mobile numbers (Figure 10.10) to receive emergency messages on their mobile devices. Once the number is entered, D needs to be pressed to exit the mode.

10.4.4 Mode C

If the user makes an incorrect entry, mode C provides access so the user can edit the error. It is therefore called "clear mode."

10.4.5 Mode D

To move to the next step or to exit a certain mode, D is pressed. It is named "enter mode."

The complete hardware model setup is shown in Figure 10.11.

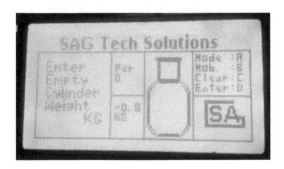

FIGURE 10.8
Activate empty cylinder weight entry in KG.

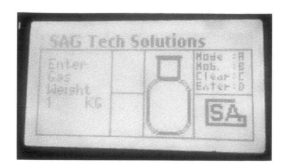

FIGURE 10.9
Mode A: activate net gas weight entry in KG.

FIGURE 10.10
Mode B: activate interface to enter mobile number.

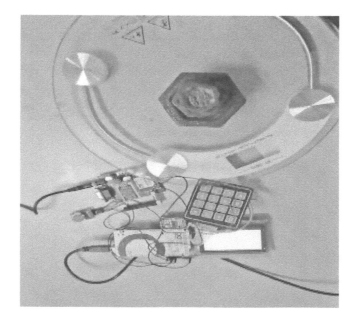

FIGURE 10.11
Complete model.

10.5 Conclusion

The proposed work yields a unique product for gas leakage detection and safety. Existing products are not capable of showing the real-time status of gas in a cylinder. The proposed prototype is cost-effective and easily accessible by every class of people. The complete setup is reliable, space saving, and feasible. An attempt has also been made to control gas leakage in case of an emergency. If no one is at home to hear the alarm and control the gas leakage, the proposed model may provide a cost-effective solution by informing the user through IoT-enabled messaging and alarming.

References

1. Nakano, S., Goto, Y., Yokosawa, K., & Tsukada, K. (2005). Hydrogen gas detection system prototype with wireless sensor networks. In Proceedings of IEEE Conference on Sensors, pp. 1–4.
2. Solis, J. L., Li, Y., & Kishs, L. B. (2005). Fluctuation-enhanced multiple-gas sensing by commercial Taguchi sensor. *IEEE Sensor Journal, 5*(6), 1338–1345.
3. Rodney Tan, H. G., Lee, C. H., & Mok, V. H. (2007). Automatic power meter reading system using GSM Net. In Proceedings of the 8th International Conference (IPEC2007), pp. 465–469.
4. Nasaruddin, N. M. B., Elamvazuthi, I., & Hanif, N. H. H. B. M. (2009). Overcoming gas detector fault alarm due to moisture. In Proceedings of IEEE Student Conference on Research and Development, pp. 426–429.
5. Shibata, A., Konishi, M., Abe, Y., Hasegawa, R., Watanabe, M., & Kamijo, H. (2009). Neuro based classification of gas leakage sounds in pipeline. In Proceedings of IEEE International Conference on Networking, Sensing and Control, pp. 298–302.
6. Mengda, Y., & Min, Z. (2010). A research of a new technique on hardware implementation of control algorithm of high-subdivision for stepper motor. In Proceedings of 5th IEEE Conference on Industrial Electronics and Applications, pp. 115–120.
7. Shinde, S., Patil, S. B., & Patil. A. J. (2012). Development of movable gas tanker leakage detection using wireless sensor net based on embedded system. *International Journal of Engineering Research and Application (IJTERA), 2*, 1180–1183.
8. Chengjun, D., Ximao, L., & Ping, D. (2011). Development on gas leak detection and location system based on wireless sensor nets. In Proceedings of 3rd International Conf. on Measuring Technology and Mechatronics Automation, pp. 1067–1070.
9. Mujawar, T. H., Bachuwar, V. D., Kasbe, M. S., Shaligram A. D., & Deshmukh, L. P. (2015). Development of wireless sensor net system for LPG gas leakage detection system. *International Journal of Scientific & Engineering Research, 6*(4), 558–563.
10. Rakesh, M., & Dagadi S. (2012). Implementation of wireless gas leakage detection system. *Proceedings of the International Conference on Sensing Technology, 6461747*, 583–588.
11. Selvapriya, C., Sathyaprabha, S., & Abdul Rahim, M. (2013). LPG leakage monitoring and multilevel alerting system. *International Journal of Engineering Sciences & Research Technology, 2*(11), 3287–3290.
12. Hema, L. K., Murugan, D., & Chitra, M. (2013). WSN based smart system for LPG detection and combustible gases. *International Journal of Emerging Trends and Technology in Computer Science*.
13. Sasikumar, C., & Manivannan, D. (2013). Gas leakage detection and monitoring based on low power microcontroller and Xbee. *International Journal of Engineering and Technology (IJET), 5*, 58–62.

14. Meenakshi Vidya, P., Abinaya, S., Geetha Rajeswari, G., & Guna, N. (2014). Automatic LPG detection and hazard prevention for home security. *Compusoft*, 1–5.

15. Naik, R. N., & Reddy, S. N., Kishore, S. N., & Reddy, K. T. K. (2016). Arduino based LPG gas monitoring and automatic cylinder booking with alert system. *IOSR Journal of Electronics and Communication Engineering*, 11(4), 6–12.

16. Sonawane, S., Borole, P. B., & Shrichippa, B. (2016). Smart home monitoring system using ECSN. *International Journal of Innovative Research in Computer and Communication Engineering*, 4(6), 10904–10909.

17. Iftekharul Mobin, M., Abid-Ar-Rafi, M., Neamul, M., & Rifat, M. (2016). An intelligent fire detection and mitigation system Safe From Fire (SFF). *International Journal of Computer Applications*, 133(6), 1–7.

18. Bhumkar, S. P., Deotare, V. V., & Babar, R. V. (2012). Accident avoidance and detection on highways. *International Journal of Engineering Trends and Technology*, 3(2), 247–252.

19. Padma Priya, K., Surekha, M., Preethi, R., Devika, T., & Dhivya, N. (2014). Smart gas cylinder using embedded system. *International Journal of Innovative Research in Electrical, Electronics, Instrumentation and Control Engineering*, 2(2), 958–962.

20. Internet of Things. Wikipedia. https://en.wikipedia.org/wiki/Internet_of_things

21. Barry, R., & Meijers, J. P. (2018). IoT connectivity comparison. *Polymorph*. https://www.polymorph.co.za/iot-connectivity-comparison-gsm-vs-lora-vs-sigfox-vs-nb-iot/

11

Ubiquitous Computing: A New Era of Computing

Shailja Gupta[1], Riya Sapra[1], and Sugandhi Midha[2]

[1]*Department of Computer Science Technology,
Manav Rachna University, Faridabad, India*

[2]*Department of Computer Science and Engineering,
Chandigarh University, Gharuan, Mohali, India*

CONTENTS

11.1 Introduction

Ubiquitous computing, abbreviated as ubicomp, comprises two words–*ubiquitous* and *computing*. The simple meaning of the term *ubiquitous* is anytime, anywhere. The idea behind ubicomp is of pervasive computing, i.e., making computers invisible to the user while providing endless support services to the user. The term *ubiquitous computing* was first used by Mark Weiser [1] in the computer science lab at Xerox's Palo Alto Research Center (PARC). He is also known as one of the fathers of ubiquitous computing. Weiser articulated that ubicomp is an ideal environment that makes use of a computer so naturally that even the user uses it without even thinking about it.

Weiser and John Seely Brown [1] define ubicomp as a calm technology; they expand on the definition by saying that the computer is an invisible servant that extends our unconscious. Ubicomp provides an environment, in which a network understands its surroundings, performs the desired function, and improves the human quality of life. Ubicomp creates a context-aware environment in which several devices are connected to the network and perform specific tasks. Little human intervention is required in the case of ubicomp.

Together, ubicomp is a paradigm that builds a universal computing environment (UCE) that works in an abstract manner. It hides the underlying technology, resources, and devices from application users, and assists the users in meeting their needs. Ubicomp focuses on hiding the complexity of computing, and offers ease of use to users for performing different daily activities. It is for this reason it is known by names such as pervasive computing, Internet of Things (IoT) and so on. Ubicomp is a power that assists with our daily routine tasks powerfully by being invisible.

The IoT has been called ubicomp because most IoT devices these days use the technology of ubicomp. Examples include the Apple Watch, Amazon Echo speaker, Amazon Echo Dot, Fitbit, electronic toll systems, smart traffic lights, self-driving cars, home automation, smart locks, Nest thermostats and doorbells, and countless others. Due to its abstraction mechanism, ubicomp has become an attractive technology in commercial applications. The goal of ubicomp is to create smart devices by having a network that is equipped with sensors. Sensors are capable enough to capture the data, process the data into meaningful information, and ultimately communicate the information to users.

A middleware layer is responsible for hiding the underlying complexity. It is the layer responsible for providing the services to the applications by masking the heterogeneity of computer architecture, operating systems, languages, and networks. To name a few, there exists different context-aware architecture, such as Aura, Cap-Net, Carima, FlexiNet, and Gaia. Any middleware architecture should support the following issues:

1. **Extensibility and flexibility:** If the environment is dependent on the middleware architecture, it should support adaptability, i.e., the ability to function irrespective of change in terminals and networks. A network should be open enough to add any number of hosts required at a particular amount of time without hampering its functioning.

2. **Interoperability:** Another main feature that middleware should support is interoperability. It should integrate heterogeneous computer architecture, network technologies, operating systems, languages, and devices. Many more network architectures are entering the market, which supports heterogeneous platforms. A software-defined network is an interoperable network that supports open communication among the different vendor devices in the network.

3. **Fault tolerance:** This refers to the reliability offered by ubicomp. This feature in middleware ensures continuous functioning of the applications in the presence of faults. A small error in the network should not bring down the entire network.

4. **Mobility:** Application users should be freely able to move from one place to another and still get a personalized service.

TABLE 11.1

Major Trends in Computing [2]

Year	Mainframe (Sales per year)	Personal Computer (Sales per year)	Ubiquitous Computing (Sales per year)
1940	0	0	0
1950	1	0	0
1960	3	0	0
1970	4	1.5	0
1980	3.9	9.5	0
1990	3.5	14	1.9
2000	1.8	16.2	12
2010	1.5	15.9	17.5

11.2 Trends and Impact in Ubiquitous Computing

Weiser announced that after mainframe personal computers, ubicomp is a trend that is growing and continuous (Table 11.1).

Advances in technology, such as personal digital assistants (PDAs), smartphones, palmtops, and notebooks, have spurred the growth of ubicomp.

11.3 Characteristics of Ubiquitous Computing

1. **Context Awareness:** Context awareness is a key characteristic of ubiquitous systems. A network of devices and sensors creates a ubicomp system that performs the desired function. A context-aware environment provides the user with the flexibility to move from one platform (place) to another. Context awareness facilitates the adaption to changing environments. Users can switch from one platform to another, and one place to another without noticing any difference in the service provided. Ubicomp shields an underlying complexity and makes feasible the anytime, anywhere service idea.

2. **Ubiquity:** Ubicomp networks are highly ubiquitous. The entire system is invisible to users, and still the users enjoy the benefits of the system. Mostly less noticeable and sometimes even invisible, ubicomp components interact with surrounding environments and provide the desired service.

3. **Networking:** Network components and sensors are connected to each other generally via wireless media. Due to this reason, they can spontaneously create networks, rather than act as components of a fixed environment. A wired service may also be provided as the situation demands.

4. **Embedding:** Ubicomp is integrated into the needs of our daily lives. It converts our objects into smart objects (e.g., smart traffic lights, Fitbits, electronic toll systems). Embedded microcomputers (MCUs) thus augment a physical object's original use and value with a new array of digital applications [3].

5. **Human-to-machine (H2M) interfaces:** Humans are unaware of the presence of ubicomp devices and machines. Minimal to no user intervention is involved. Users enjoy the services without actually noticing its existence.

11.4 Technological Foundation of Ubiquitous Computing

Advancements in technology have led to a decline in the cost of hardware and devices, such as microprocessors that are further equipped with sensors and other wireless communication facilities. Advancements have also given rise to information processing computing that is ubiquitous and prevails today in almost every object used in our daily lives.

The following features characterize the technological foundations of ubiquitous computing:

1. Ambient intelligence into everyday objects.
2. Decentralization of devices/systems and their comprehensive wireless network connection.
3. Anywhere, anytime service.
4. Flexibility and adaptability to changing environments.
5. Hiding complexity, making use of abstraction, which further simplifies user experiences.

The driving force behind the technological improvements has proven the prediction of Gordon Moore [1]. He stated in his Moore's law that computing power available on microchips doubles every 18 months. This growth is clearly visible in modern processors, microchips, storage capacity, and network bandwidths. Recent developments in the field have given birth to embedded sensors in everyday objects (Figure 11.1). This enormous growth in technology has led to the development of very small computers, known as "information appliances," that are connected via wireless frequencies.

Weiser [1] sees technology only as an end product, i.e., the human-centric services it offers to users. Users do not even realize they are using ubicomp technology. Varieties of ubicomp are quickly gaining popularity.

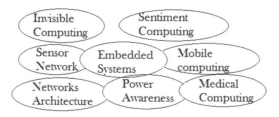

FIGURE 11.1
Ubiquitous computing technological foundations [4].

11.5 Drivers and Hurdles of Ubiquitous Computing

The major issues in ubicomp include:

1. **Interoperability:** The big obstacle is that every system or software must be able to understand every other device in the network. It becomes a challenging task to ensure that all devices in the network can freely communicate with each other. Devices in the network are from different vendors, and making the communication flow open and feasible is difficult.

2. **Reliability:** Devices must achieve high levels of reliability, especially when the devices are ad hoc coupled to each other to achieve the desired objectives. Reliability becomes a bottleneck in the development of ubicomp. A connection failure among devices is a major bottleneck while setting up the connection.

3. **Standardization:** Ad hoc coupled networks lack certain standards; however, some type of standardization is of utmost importance as the functionality interconnection of individual devices is a key feature of ubicomp. Devices from different vendors follow different protocols, so establishing a common standard among those is a tedious task.

4. **Resource discovery:** The capability of devices to define their behavior in an ad hoc network is a key requirement. No device in a ubicomp network has prior knowledge about the function of all other devices in the network.

5. **Network:** Different devices are prevailing in the ubicomp network. It is a big task to establish a network that is invisible to users, while still allowing users to enjoy its benefits. A ubicomp network is a connection of ad hoc, unaware components.

6. **Privacy and security:** Because ubicomp devices are equipped with sensors, privacy and security [5] become a key issue. Users get worried about their privacy as their actions and gestures may be monitored.

11.6. Impact of Ubiquitous Computing

The driving force behind ubicomp is to make computers invisible to their users but also natural, friendly, and fitted to the environment. The name itself implies everywhere, anywhere. In this section, we discuss the impact of ubicomp on the economy, privacy, and social lives.

11.6.1 Impact on Privacy

Ubicomp involves interactions across multiple organization boundaries. Due to this reason, it requires "out of the box" setup for security. User actions are monitored, which demands implementation of context-aware adaptive security. Users are always worried that security mechanisms will be unable to differentiate between authentication and anonymity. A user should at least possess control on information generated about him/her. Different devices generate different information. Every device can communicate with others, but still reliability of information cannot be guaranteed.

11.6.2 Economic Impact

This area tracks how business processes can improve, and how ubicomp can help to create completely new business models. Ubicomp is a step further in the electronic commerce process.

11.6.3 Social Impact

Ubicomp certainly has social impacts on society. As a device gets inputs from the surrounding environment, the question related to a device's social implications becomes obvious, such as: Does it support privacy? Will it be a social solution? Is there an ethical solution to protect privacy?

11.7 Ubiquitous Computing Security

Security and trust in technology is important. In ubicomp, trust [6] refers to two terms: (1) reliability trust, and (2) decision trust. Ubicomp no doubt offers numerous benefits to individuals and society by embedding technology in their everyday objects. Yet, this exposes significant risks, such as information disclosure and lack of centralized control. These risks come under the umbrella term *privacy*. Designing privacy becomes an inherently important and complex task.

The following security and privacy challenges relate to ubicomp. All the security aspects must be considered to build user trust in the application.

1. Nothing is fixed; ad hoc network
2. Decentralized control; no central management
3. Each entity can interact with other entity neighbors; no trust checking
4. No knowledge of information sources
5. Unreliable information is prevailing

11.8 Working of Ubiquitous Computing

Ubicomp has become possible with the integration of sensors in day-to-day life that lead to communication among the components. In Figure 11.2, the bottom layer, also called the physical layer, compromises various sensors. These sensors are attached, integrated, worn, or carried by people, industries, campus, vehicles, fields, and smart devices. Sensors come installed in devices these days, such as GPS, accelerometers, and so on.

The layer above the physical layer is the infrastructure platform or wireless platform that uses 802.11 group of networks, which ensures lower latency. Other technologies, like ZigBee and near-field communication (NCF) provide low-cost alternatives to connect multiple devices by connecting parent sensors to different child sensors that are wirelessly controlled.

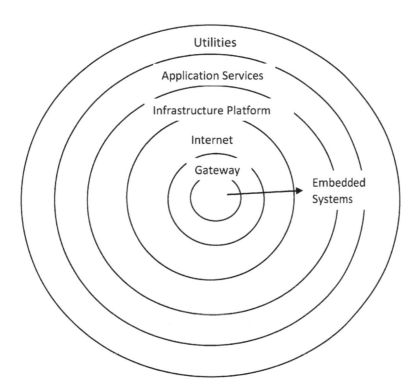

FIGURE 11.2
Ubiquitous computing stack [1].

The next layer above it includes application services. The data are collected, mined, and examined for all the patterns. Patterns allow the smart applications to make modifications via devices, like tablets, smartphone or smart devices, handheld devices, and so on.

For example, if the user is a cardiac patient, then wearing a tiny device will monitor the health of the patient by generating an alert to the patient doctor's mobile device. This scenario is illustrated in Figure 11.3.

In order to generate a ubiquitous environment, various components, such as web, smart devices, software, and communication frameworks collectively work to process and model the information using different materials, protocols, environments, and products.

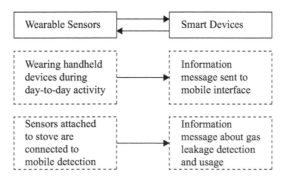

FIGURE 11.3
Smart device interaction [7] example.

11.9 Anticipating Ubiquitous Computing

Today's world is filled with tech-savvy people who are digital-literate and do not hesitate in experimenting with technology. They are comfortable with working on multiple gadgets and wish to be connected to the world using these devices. Table 11.2 shows the role and affect of smart devices in different domains. In industries such as gas, energy, and water, sensors can predict supply and demand usage and analyze usage data. The sensors send the data on Internet-connected smart devices to provide cost savings on resources.

In manufacturing, sensors are used for anomaly detection in equipment functioning, and automatic quality monitoring in the production line. The sensors, like the ones that can be placed on equipment, wearable sensors, microphones, and cameras, can be installed on a central console. Connectivity is provided using mobile devices for remote consultancy. Following are the outcomes of using smart devices in a manufacturing line:

- Quality and sustainability improvement
- Emission detection and anomaly detection
- Production line scheduling for analyzing productivity of different divisions, power rates, and worker availability

Healthcare: In healthcare, medical sensors are widely used for historical correlation and anomaly detection. Wearable devices also help patients by monitoring patients' vitals continuously and by providing faster response to the patients. The patients are provided teleconsultation using super specialist medical computers with the help of support systems and clinical guidance.

Insurance: Sensors that monitor the usage and condition of insured entities help insurance companies by providing usage reports of entities. These sensors help in detection of an anomaly and provide detection in the usage patterns. Internet-enabled mobile devices can be connected with the sensors to provide usage-based insurance to users.

Banking: In banking, biometric sensors and NFCs provide analytical reports on financial data. They help is automating secured transactions or payment systems on devices using

TABLE 11.2

The Hierarchy from Mainframe to Ubiquitous Computing

Mainframe	Many consumers, one computer
	Computer used to take up the entire room
Personal computer	One consumer, one computer
	General purpose input-output device
	All jobs done on a single device
	Human-to-computer Interaction
Ubiquitous computing	One consumer, many computers
	Size and shape of the computer is changed and can be hidden
	Minimum human intervention
	Context-aware
	Devices connected to network to perform tasks
	Computer-to-human interaction

the Internet. The automation of processes leads to fast, secure, convenient, and contactless transactions via mobile devices.

Consumer Industry: Smart devices play an important role in the consumer industry by recording viewership data. They automate the process by extracting context from sensor data and provide real-time user profiling. The industry can make decisions on its products by:

- Extracting recommendations from users' mobile devices
- Incorporating suggestions that users provide as automated feedback
- Providing advertisements for the products based on usage
- Developing or coming up with new product ideas based on user likings and usage habits
- Generating consumer applications such as alerts for consumer products to help the end customer

Education: In education, smart devices, such as camera, audio, and video sensors are widely used. They provide analytics of video and audio that help in providing interactive distance education via applications on tablets or other Internet-enabled devices. They also provide interactivity on devices, like tablets, kiosks, and television.

Telecom: Sensors that are used for mobile-to-mobile communication provide industry vertical specific analytics. These sensors are available on almost all mobile devices that are connected via the Internet.

11.10 Application Areas

Ubicomp follows the concept of everything-everywhere, i.e., to connect all the areas of work where computing plays any role. It therefore leads to ubiquitous information, data, and knowledge. With the Internet available to all devices, like smartphones and handheld computers, and accessibility of applications, like emails, downloading, uploading, and others, the exchange of information has become easy. Ubicomp facilitates these processes by providing a new dimension to exchange information, data, and knowledge. Ubicomp helps in various applications by running a lot of processes in the background, thereby giving a feeling of automatic computing. The speed of computation depends on the following applications [8].

1. **Communications:** Communication via exchange and transmission of data, knowledge, and information is necessary in all the fields. It is a prerequisite for every information technology area.

2. **Logistics:** Automating and optimizing the existing model of logistics in the form of tracking and disposal of raw material as well as semifinished and finished goods are required to close the existing gap in IT systems (control systems).

3. **Motor traffic:** Modern vehicles are equipped with assistant systems for navigation that support the driver instantly and spontaneously. Networking different vehicles with each other and including systems that can record the behavior of an automobile and can map to different locations are the systems of the future.

4. **Military:** Developing systems that can prevent and fight external threats is one requirement of the military. It also includes the development of new weaponry systems, as well as systems for collection of data and processing of information acquired by the data.

5. **Production:** In automatic production systems, the flow of raw material, processing, packaging, and transport are handled by machines. Ubicomp will further help to ease the functionality of decentralized systems for production in smart factories by automatic monitoring, configuring, and controlling the equipment and machinery.

6. **Smart homes:** In smart homes, the smart objects are available for necessities, such as lighting, heating, communication, and ventilation. These technologies are at our fingertips in smart homes that automatically control and adjust to the needs of family members.

7. **E-commerce:** Ubiquitous smart objects allow scaling the services according to traffic needs or demands of the customers. They also help in maintaining the speed for e-commerce websites and providing the facility to pay for the services required. They provide redundancy-tolerant systems that help keep the data backed up, secure, and easily available.

8. **Inner security:** Smartcards, e-passports, and so on, are some of the ID system applications that are making inner security computing ubiquitous. Monitoring systems will be of utmost importance as we move into the future, where applications can vary from environmental conservation to surveillance.

9. **Medical technology:** Employment of tiny, self-sufficient, feature-rich, and connected medical devices in ubicomp enables round-the-clock oversight on the well-being of the elderly (from the comfort of their homes) and artificial intelligence-driven implants.

11.11 Ubiquitous Technologies

Ubiquitous computing cannot be performed using a single technology. For example, an object can be implemented as a smart object or it can be created using different components. For this reason, different technologies [9] are responsible for ubicomp, but some characteristics can be true for a variety of application areas. For example, mobility, ad hoc networks, context awareness, and autonomy are important characteristics for communication technology.

1. **Microelectronics:** Microelectronics that deal with the development and manufacture of micro-integrated circuits have become necessary components in a large number of devices. Consumer electronics, medical technology, and the automotive industry depend on microelectronics for day-to-day needs. Institutions have been working on reducing chip size, reducing cost, increasing integration density and functionality, and many more features. Microelectronics are mature and available widely. They also do not create any bottlenecks for computing.

2. **Power Supply:** In order to operate, electronic devices require a power supply. With improvements in chip technologies system size, power demands have been reduced and performance levels have improved. Still, establishing a constant power supply remains an issue of concern. For power supply, some devices operate on energy from the electric grid, while others, like wireless systems or mobile devices, depend on stored power sources. Therefore, power sources are required that provide long-term stability and excellent reliability. Moreover, improvements in power supply have allowed generating power without batteries. Some examples are photovoltaic generators (using solar cells, piezoelectric generators (mechanical to electrical energy), thermoelectric generators (heat to electrical energy), electromagnetic generators (based on dynamo principle), capacitive and electrostatic generator (uses electrostatic charges), and thermomechanical generators (mechanical to electric energy).

3. **Sensor Technology:** A sensor is an electronic component that registers the characteristics of its surroundings and processes, amplifies, and converts the information to digital signals. Instead of using sensors as a separate entity, sensors are integrated in chips to create a sensor element. The major challenges faced in this industry are the size and weight of the sensors; integration, reliability and performance of the sensors; decreasing power consumption; and generating low-cost systems. Sensor technology is considered a technology that is well-established and does not possess any technical barriers.

4. **Communication Technology:** Information communication technology (ICT) is communication between objects; ICT helps drive ubicomp. In telecommunications, like telephones, satellite communications, or mobile communications, the important subfields are radio engineering, technical informatics, signal transmission technology, switching technology, communications engineering, microelectronics, and communications networks. A large number of standards, like 802 family, are devised for communication. Many mobile telephone providers and digital device suppliers are now working with standard organizations to develop solutions for information and communication as they recognize ICT as an important key to UC.

5. **Localization Technology:** Equipping smart objects with transmitters and receivers that provide precise localization has always been an important concern for ubicomp. Location-based services using localization technology has become an absolute necessity. The three important types of localization systems are satellite-supported systems (GPS) used for navigation and localization, indoor localization systems, and cellular supported systems.

6. **Machine-to-Machine Communication:** A ubiquitous system acts as a distributed system with lots of components working at high speed. Therefore, it becomes vital to provide a standardizing system to machine interfaces. Service-oriented architecture (SOA) has the uneasy target of streamlining application integration. It is based on the model of sharing distributed systems (reusable) across applications. Open standards supporting usage of IP-based web services, a platform for securely transporting messages (enterprise service bus [ESB]), and a dedicated integration box combine to form the basic principles of SOA.

7. **Human-Machine Interface:** With the expansion of technology and invention of smart objects, the need to compute data automatically with little or no human

intervention has become an important requirement. Such smart objects require an interface to communicate between the users and applications, also referred to as a human-machine interface. Such an interface enhances the interactions with the user by including visualization components over monitor-keyboard principle. Many industries are working intensively on these interfaces, such as consumer electronics, computer systems, healthcare technologies, and automobile industries.

In general, no model exists for developing a human-machine interface. The human-machine interface semantic model is therefore required to be developed for generation of real-world understanding. Many experts believe that the evolution of this interface will lead to evolution in ubicomp. Research in this area has led to development of speech technologies, like mobile assistants, fingerprint sensing interfaces for smart locks, health-tracking interfaces like Fitbit, and many more.

11.12 Use Cases of Ubiquitous Computing

1. **Smart Remote Control:** A context-aware smart television remote control can detect the user by his or her way of handling and using the remote. It also learns the user's choices of channels and provides preferences for the same.

2. **Vacation Planning Application:** A smart application can take information such as location, calendar, and user preferences, from an Internet-enabled mobile phone, to provide vacation and travel suggestions to the mobile user.

3. **Apple Watch/Smart Watch:** Smart watches are a perfect example of ubicomp where the watch informs the user about the phone call, messages, and other notifications.

4. **Amazon's Echo Devices:** This is another important example of ubicomp in which the device plays songs after the user gives song requests.

5. **Nest's Smart Thermostats:** A thermostat is an electronic device that senses the temperature and performs actions accordingly. Nest's smart thermostat is a self-learning and programmable thermostat, which turns off when there is no one at home, and can be turned on from anywhere using a mobile application. It is a proven energy-saving technology.

6. **Mimo Infant Monitoring:** Wearable on a baby's T-shirt, this breathing-monitoring device is connected to a mobile application that helps to monitor a baby's sleep and breathing patterns. It also alerts the parents whenever there is a change in the pattern of breathing or activity of baby.

7. **Proteus Discover:** This is a wearable electronic patch with ingestible sensors and derives patient health patterns to improve the effectiveness of medication treatment. It connects to a mobile application to get the insights of the collected data from the sensors to create informed healthcare.

8. **LIFX Smart Lights:** These Wi-Fi-enabled lights can be controlled or automated with the LIFX mobile app or voice assistants. They come as multicolor lights; the color of the lights can be changed with the mobile application.

9. **SmartPile System:** This is a sensor technology-based device to access the effectiveness of concrete material used in the piles of buildings. The SmartPile system connects to the software to report real-time health and capacity information of the piles of buildings. The embedded sensors provide highly accurate insights of the static as well as dynamic health of the piles, which can be used for accessing the strength of the building as a whole.

10. **Eyedro Electricity Monitor:** Eyedro's electricity monitor gives real-time electricity usage details so that the electricity bill at the month's end does not surprise you. It is an affordable way for homeowners to take actionable measures, if any are required.

11. **OneFarm:** OneFarm is a farm management tool that is used to collect data from the fields using sensors, and helps farmers to improve field profitability. It is also optimizes farming operations by using a centralized platform for collecting and analyzing the field's data in order to guide decisions to improve the quantity and quality of crops.

12. **Radio frequency identification (RFID):** RFID [10] is an amazing example of ubicomp where data are encoded in the form of smart labels, called RFID tags, which are captured by machines via radio waves. It is sometimes similar to barcoding in that the data are saved in a database in both technologies. The difference is that RFID can be read out of the line of sight, but barcodes must be aligned with an optical scanner. A very common use of RFID tags is at toll plazas. Regular commuters pay the toll in advance and get an RFID tag for their vehicles, so they need not stop at the toll plaza to pay the toll tax. The machine scans the RFID tag mounted on the vehicle and automatically opens the gate without any manual intervention. Other tasks that can be performed using RFID are:

    * ID badging
    * Controlling access to restricted areas
    * Inventory management
    * Supply chain management
    * Personnel tracking
    * Asset tracking

13. **Self-Driving Cars:** Self-driving cars, also known as driverless or autonomous cars, are cars that are capable of sensing the environment around and moving without any human input. In order to know their surroundings, these cars use a mix of sensors such as GPS, radar, sonar, computer vision, odometry, LiDAR, and inertial measurement units. The sensory information is input to the advanced control systems, which interpret the information to identify the appropriate path or obstacles.

11.13 Security in Ubiquitous Computing

Ubicomp allows connecting and computing on the move. It relies on small devices, called sensors, embedded in our everyday objects to collect information and communicate and deliver it to other objects wirelessly, for example, RFID in cars and automatic temperature controllers in an air conditioner. It is called a *calm* technology as it works without the obstruction of humans. As technology becomes more advanced and embedded, new devices become more and more intelligent. With the advancement of technology, device-to-device communication has increased manyfold, which has given birth to smart homes, smart mobile phones, smart medical appliances, smart watches, and much more.

Ubicomp is all about the interconnectedness of hardware and software such that no one even notices its presence. It is changing our everyday life, and will also impact our socioeconomic lives in a positive as well as negative manner. Industries relate it to omnipresent information processing technology and pervasive computing as they look for its use in the field of electrical devices, toys, tools, household appliances, and so on, to digitize everything. Providing computing anywhere and anytime brings many advantages and a few sets of challenges as well. Safety and privacy are also affected both positively and negatively. Following are the security issues that are affected:

1. **Privacy issues:** Ubicomp deals more with machine-to-machine interactions where all the operations need to conform to the data protection standards. This acts as the first requirement of the network so that privacy of data is maintained and the devices while communicating are protected from outside attacks, like man-in-the-middle attack, access network attack, and so on. A secure and careful architecture is required to ensure the privacy of content shared among the devices. The main reasons that the privacy is hindered is because potential victims do not care about it and due to personal interest, victims try to violate it [11].

2. **Trust issues:** Trust is the agreement between two parties for secure, reliable, and trustworthy communication or interaction between them. The devices need to maintain trust agreements while sharing their information between them so the correct information reaches correct hands only, and there is no information leakage. The data among devices are shared wirelessly with or without the involvement of service providers. When there is no involvement of a service provider, the devices only need to conform to the data protection standards. But, when the data are sent through a service provider, the data protection standards need to be designed for the service providers and trust needs to be maintained regarding the safety of personal data.

3. **Social issues:** Social values and terms are very important while making data protection and sharing standards for security, trust, and privacy of data and systems as a whole. Also, the manner in which the devices have to be connected and share information can be different for different individuals or societies. The extent of information needed to be shared among devices can also vary. This as a whole can affect one's perception or mindset towards the technology. Also, the frequent use of ubicomp can also lead to unintended handling of devices where the devices are communicating with each other without the knowledge and intervention of humans. This can lead to accidental sharing of information to the wrong devices or people.

Many mechanisms have been created to combat the above-mentioned security attacks. Some include masking mechanisms, proximity detection schemes, game-based approaches, consent and notification approaches, negotiation approaches, obfuscation, and blocker tags for RFID-based systems [12].

11.14 Security Attacks in Ubiquitous Computing Networks

With the growth of technology, security has become a popular topic of research. Confidentiality, integrity, information availability, and real-time sharing are some common areas where a sound security protocol is required [13]. There have been various security attacks in ubiquitous environments, as illustrated below:

1. **Man-in-the-middle attack:** A man-in-the-middle (MITM) attack is a security attack when someone eavesdrops or positions himself or herself between the conversation of a user and an application, impersonating one of them making it appear as a normal conversation. The basic goal of this attack is to get some personal information, say bank account details, credit card information or login credentials.

   Target areas for such types of attack include e-commerce sites, software-as-a-service (SaaS) businesses, and financial applications. This information can be used to user identity theft, unethical funds transfer, password change, and so on. It is somewhat similar to a postal worker who opens your bank statement without informing you, saves your account details, closes the statement, reseals the envelope, and delivers it to your door.

2. **Access network attack:** An access network attack is a form of unauthorized computer resources access. This access can be done physically by accessing the computer resource or remotely by gaining entry through a network. An access network attack can be done by an outside individual or a group who gains access through a network to steal confidential information. Access network is also possible by an inside individual entering in network areas where he/she is unauthorized in order to hack some information. Some types of access attacks include:

   a. **Password attack**: A password attack is using tools to guess the password of a user and then hacking the information. Every individual login and logout to shared resources on servers or routers uses authentication passwords; the attacker repeatedly tries to log in to gain unauthorized access to resources using various software tools such as Cain and Abel or L0phtCrack. These password-recovery tools use dictionary words to predict the passwords.

   b. **Trust exploitation attack**: A trust exploitation attack compromises a trusted host to use it to attack other hosts in the same network. For example, consider a host in an organization network that is protected by firewall (Figure 11.4).

      This host acts as an inside host that is accessible to the trusted host outside of the firewall, known as the outside host. That is, the inside host is accessible to the outside host, which makes the inside host vulnerable to attack by the outside host. The hacker in this case can hack the outside host and gains access to the inside host to steal unauthorized data. Trust exploitation attacks can only be controlled by implementing strict protocols on the trust levels of hosts within the network. Hosts outside the firewall should never be trusted by the hosts inside the firewall. Communication between the inside and outside host must be authenticated via passwords or agreements.

   c. **Port redirection**: Port redirection is another form of trust exploitation. The attacker uses the outside host, sometimes called comprised host, to get access

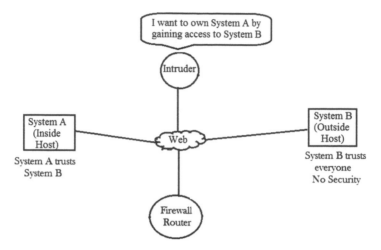

FIGURE 11.4
Trust exploitation attack [14].

through the firewall and access the inside host (Figure 11.5). The attacker can also install software on the outside host system to redirect its traffic from the inside host directly to him/her. Netcat is one such tool that provides this functionality.

3. **Illegal connection attack:** Generally, household appliances that are connected through a home gateway to multiple networks are managed or controlled via some web-based application. The attacker can hack the home gateway to control the web-based application, thereby pretending to be the administrator of the network of appliances. In this way, he/she will get the control over the working of the appliances, and hence can also steal some important information and misuse it.

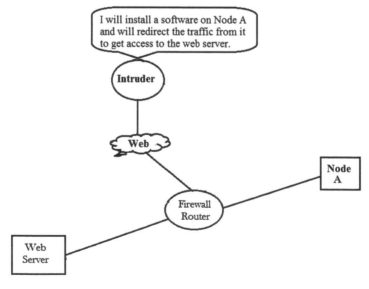

FIGURE 11.5
Port redirection attack [15].

TABLE 11.3

Major Players in the Market

Apple	Apple watch that tracks user's motion, activity, elevation, heart rate, etc.
Cisco	Provides B2B solutions for large-scale smart home automation
Microsoft	Microsoft Band tracks all fitness-related metrics and moon shots like HoloLens are building cutting-edge mixed-reality experiences
Amazon	Amazon's Echo product line provides a vast variety of functionalities—the Alexa-enabled echo speakers, smart microwave, echo input, dash button, etc.
Facebook	Oculus VR is the leading virtual-reality headset with the ability to create both augmented and virtual-reality experiences
Google	Google's Nest product line consists of smart thermostats, surveillance cameras, video doorbells, door locks, smoke detectors, etc.
Beddit	Provides smart bedding that monitor sleeping patterns, breathing, heart rate, environment, and movements
Tesla	Tesla car learns the driving habits and provides intelligent solutions to the car by automatically adjusting temperature and other conditions to the needs of the driver
Hudway	Turns the windshield into a jet fighter cockpit

4. **Capturing sensitive data:** Ubiquitous systems use electronic sensors, which have very poor computational power of monitoring data or devices. The attacker can very easily keep a receiver near the sensor and get the sensitive information directly from it. This information leaked directly from the sensor can be misused or manipulated.

5. **Denial of service (DOS):** Denial of service [16] is a cyber attack that shuts down the network or a machine temporarily or indefinitely so as to make it unavailable to its users. This is done either by flooding the target machine or network with traffic or sending superfluous requests, thereby preventing legitimate requests to be noticed. It is like crowding the door of a shop and making it difficult for the real customers to enter the shop. Such attacks are more common to media companies, trade organizations, and banks. The goal of such an attack is to flood unnecessary traffic, causing the server to slow down, crash, or stop. These attacks may not result in any type of theft of information or assets but will definitely cost a great amount of time and money to recover and restore services.

11.15 Major Players and Current Trends in the Market

Table 11.3 presents the major players and current trends in the market.

11.16 Conclusion

Ubiquitous computing integrates into everyday objects to network them. Ubicomp is being adopted by all types of computing technologies to bring together the power of different computing machines and provide a hassle-free experience to the users.

The technology is setting the expectation levels of users and industry very high in terms of tasks being done automatically by the machines. The application base of ubicomp is expanding at a fast pace. The use of RFID on vehicles has reached a high maturity level, and now RFID is being used in other applications, like supply chain management and inventory management. Mobile applications and smartphones have paved the way for new devices, like smart TVs, smart watches, and so on. These devices are revolutionizing the ways of machine-to-machine communication and setting the levels high. The technology is very user-friendly, and can be adopted by everyone, but some technical aspects still require research and development for being adopted by other industries, like banking and financial sectors.

The major concerns of ubicomp are privacy and trust, which act as barriers to its long-term success. Methods and techniques need to be researched to make safe and dependable systems that have the ability to diagnose errors. This is important because ubicomp systems need to gain user trust in order to be adopted in critical areas where required safety levels are very high, such as payments and healthcare.

In the time to come, ubicomp will change working in the personal, public, and economic realms. This brings new functions and more opportunities to the objects to communicate anywhere and anytime. Ubicomp will also change the present centralized scenario of the transportation, service, and public sector. As a whole, the technology will bring amazing changes to personal lives as well as businesses and industries.

References

1. Jaydip, S. (2012). Ubiquitous computing: Applications, challenges and future trends. In *Embedded Technology and Wireless Technology* (pp. 1–40). Boca Raton, FL: CRC Press Taylor & Francis Group.
2. Sassone, V. (2008). Elements of foundations for ubiquitous computing. University of Southampton, Electronics and Computer Science. https://eprints.soton.ac.uk/268623/1/inriaPres.pdf
3. Man in middle (MITM) attack. Web application security center. (2018). https://www.incapsula.com/web-application-security/man-in-the-middle-mitm.html
4. TCS. (2018). Ubiquitous computing: Beyond mobility: Everywhere and everything. https://sites.tcs.com/insights/perspectives/enterprise-mobility-ubiquitous-computing-beyond-mobility
5. Matzner, T., (2014). Why privacy is not enough privacy in the context of "ubiquitous computing" and "big data." *Journal of Information, Communication and Ethics in Society*, 12(2), 93–106.
6. Denko, M. K., Sun, T, & Woungang, I. (2011). Trust management in ubiquitous computing: A Bayesian approach. *Computer Communications*, 34(3), 398–406.
7. Harris, T. (2018). Ubiquitous computing—Living in a smart world. https://www.thbs.com/blog/ubiquitous-computing-living-in-a-smart-world
8. Jeon, N., Leem, C. S., Kim, M. H., & Shin, H. G. (2007). A taxonomy of ubiquitous computing applications. *Wireless Personal Communications*, 43(4), 1229–1239.
9. Schoch, T., & Strassner, M. (2002). Today's impact of ubiquitous computing on business processes. University of St. Gallen.
10. Sheffi, Y. (2004). RFID and the innovation cycle. *The International Journal of Logistics Management*, 15(1), 1–10.
11. Stajano, F. (2009). Security issues in ubiquitous computing. *Handbook of Ambient Intelligence and Smart Environments*, 281–314.

12. Kusen, E., & Strembeck, M. (2016). A decade of security research in ubiquitous computing: Results of a systematic literature review. *International Journal of Pervasive Computing and Communications*, 12(2), 216–259.

13. Hyatt, Z., Reeve, J., & Boutle, C. (2007). Ubiquitous security for ubiquitous computing. *Information Security for Ubiquitous Computing*, 12(3), 172–178.

14. Orbitco. What network trust exploitation attack? Orbit-computer-solutions.com. http://www.orbit-computer-solutions.com/type-of-network-attack-trust-exploitation/

15. Orbitco. Port redirection attack—Explanation with examples. Orbit-computer-solutions.com. http://www.orbit-computer-solutions.com/port-redirection-attack/

16. Lee, S., Kim, D. S., Lee, J. H., & Park, J. S. (2012). Detection of DDoS attacks using optimized traffic matrix. *Computers & Mathematics with Applications*, 63(2), 501–510.

12

Detection of Moving Human in Vision-Based Smart Surveillance under Cluttered Background: An Application of Internet of Things

Dileep Kumar Yadav

Dept. of CSE, Galgotias University, Greater Noida, India

CONTENTS

12.1 Introduction

Detecting moving objects in video surveillance systems is a wide progressive research area for security purposes in private and public areas of daily life [1–3]. It is useful for many real-time image analysis applications, such as defense, navy, indoor-outdoor visual surveillance systems, human-machine interactions, traffic analysis, robotics, and surveillance for restricted zones. A visual surveillance system is a challenging technology that faces many problems due to environmental effects and noise due to illumination variations; movement of tree leaves, banners, and flags; and fountains [1–5]. The goal is to develop an effective scheme for detecting moving objects and also to resolve the problems associated with the background subtraction technique. Some key statements are given as to how to develop new methods by improving the existing methods to compute the threshold that could classify pixels more effectively, and give better performance with the other existing methods.

In a video surveillance system, the camera is mainly static with a dynamic background that gives an idea of moving object detection from a video [5–9]. Moving object detection can be performed with several techniques, such as the background subtraction technique, region-based image segmentation, temporal differencing method, active contour models method, and optical flow method. Background subtraction is a popular approach, which is suitable for the static camera and dynamic background that leads to detection of the moving object in the video by subtracting the current frame from the background reference frame [10–13]. The moving object can be detected by comparing the current video frames with the background frame. The camera could be fixed and the background could be dynamic [8, 14–17]. The background model requires an adaptive behavior for the changes with dynamic backgrounds.

Moving object detection of pixels in every frame is determined by the statistical parameters that use the background subtraction method [3, 11, 17–20]. Background subtraction consists of two phases, as given below.

Training Phase: In this phase, the background modeling has been done to evaluate the trimmed mean, which is based on the simple average method.

Testing Phase: In this phase, the threshold values are generated to evaluate the difference of the current frame (F_i) and background reference frame (B_R), and then pixels are classified using the difference frame that is based on the varying threshold value.

Threshold is a well-known approach in image segmentation that reduces the color image into the binary image. The main goal of the threshold is to elicit each pixel from every frame that represents a range of intensity, even if the information is binary (Figure 12.1).

So, the objective of binarization is to distinguish the pixel that belongs to the background area with the different intensities and the true foreground area with a single intensity. There are two types of threshold algorithms, as given below.

1. **Global threshold:** In this threshold, a single threshold value is used for each pixel in every frame. When the values of pixels of all components and the background are uniform in their respective values over the whole image, a global threshold can be used.

2. **Local or adaptive threshold:** In local threshold, different threshold values are used for the different local regions.

In an IoT cloud-oriented video surveillance environment, the visual surveillance is the detection of moving objects from a scene that is captured by closed-circuit television (CCTV), Internet protocol (IP), etc. (Figure 12.2). Surveillance security areas include indoor-outdoor, banks, shopping malls, transportation, agriculture, restricted zones, border, defense, army, games, and so on [6, 7, 21, 31, 32]. It can be applied in behavior or activity analyses or sports and event detection/tracking.

FIGURE 12.1
Process of generation of threshold.

This chapter presents critical challenges with the application areas and suggests an overview of frameworks through Internet of Things in the cloud (IoTC) environment. This work also describes the benefits, risks, and IoTC-based video surveillance, which is based on suitability for robotics, transportation, healthcare, navy, army, navigation, social media, and other areas. Due to technological advancements and new research, video surveillance is moving towards the IoTC environment, where billions of devices have the capability to collect and transmit video data in terms of frames via the Internet

IoT devices allow network cameras to think and analyze independently. These artificial intelligence (AI)-based devices are capable of making smart decisions themselves. For example, a mesh of network cameras may communicate and generate an alert message for the next camera about an object entering the scene from the left side or right side of the scene. IoT-based devices may also cover up for the damaged or obstructed objects in a scene. IoT-enabled devices are very helpful for healthcare domains using video surveillance systems.

12.2 Related Vision Techniques

Various approaches are available for moving object detection in videos. In this chapter, some previous approaches and algorithms are discussed using background subtraction. Wren et al. [2] developed the background model independently at each pixel of every frame using the probability density function (PDF) on each previous pixel value. Stauffer et al. [6] investigated a Gaussian mixture model (GMM)-based method that is modeled individually using three to five Gaussian mixtures in a consecutive manner. This method detected many false-positive alarms, so many researchers aim to enhance this technique further.

Lee et al. [14] improved the Stauffer et al. method by implementing a wiener filter and adaptive learning rate in which the convergence rate improved without compromising model stability. Haque et al. [7] improved both the Stauffer and Lee methods and also developed the Gaussian mixture model, which is based on the background subtraction technique. Jung [9] has developed a trimmed mean method based on the background model using statistical signifiers. Various literature has discussed different issues [22–25]. Various methods are used for detection of moving objects. Some popular methods are as follows:

- **Phase-correlation method:** This method is simply the inverse of the normalized cross-power spectrum in video frame.
- **Frame difference method:** This method demonstrates the absolute difference between two consecutive frames. This method generates poor information in cluttered environments.
- **Block-based method:** This method simply maximizes the normalized cross-correlation or minimizes the sum of squared or absolute differences.
- **Discrete optimization method:** This method quantized the search space then a label assignment is applied at every pixel for frame matching. Through this, the deformation between the source and target pixel is minimized [21]. The optimal solution is evaluated through linear programming, max-flow min-cut algorithms, or belief propagation methods.

- **Differential method:** The differential method normally computes the movement by a moving pixel in every unit of time. There are various differential methods, such as the Lucas-Kanade method, Horn-Schunck method, Black-Jepson method, and Buxton-Buxton method.
- **Background subtraction method:** This method [6, 7, 14, 26–28] simply computes the absolute difference between current frame and reference frame (background model). Here, initially we model the background using few initial frames then compute the difference by subtracting it from the current frame. This method is explored in detail in this chapter.

12.3 Proposed Work

In the proposed method, moving object detection of each pixel in every frame can be obtained by the statistical parameter using the background subtraction method [26, 27, 29, 30]. The proposed work of background subtraction consists of three phases, as described below [6–8, 22, 28–30].

12.3.1 Training Phase

In this phase, the background modeling has been computed using the simple average method. The threshold has been calculated by the standard deviation, as shown below.

- **Background modeling:** In the training phase, we have constructed a model, that is, a background reference frame using the initial few frames. This concept is inherited from the [21, 24, 25]. The background model is constructed, as given below.

$$B_R(x,y) = B_R(x,y) + \frac{F_{BG}^b}{N_{BG}^b}(x,y) \tag{12.1}$$

Where, $B_R(x, y)$ is the intensity value of a pixel at location (x, y) in background model. $F_{BG}(x, y)$ is the intensity value of a pixel at location (x, y). N_{BG} is the total number of frames for the background model construction where BG is considered as $b = 1$ to 200.

- **Threshold generation:** The threshold value is calculated using the standard deviation.

$$\sigma = \sqrt{\frac{1}{N} \sum_{i=1}^{N} (x_i - \mu)^2} \tag{12.2}$$

Where,
 σ is the standard deviation;
 N depicts about the number of values in the population;
 x_i depicts each value of the population; and
 μ shows the population mean.

12.3.2 Testing Phase

In this phase, the difference of the current frame from the background reference frame has been evaluated. The pixel classification has been done by comparing the threshold in terms of background frame and object frame, and updated the learning rate.

- **Background subtraction:** After initial background modeling, we have read the current frame (F_i) and subtracted it from the background reference frame B_R and the difference frame, as shown in Eq. (12.3).

$$D(x,y) = |F_i(x,y) - B_R(x,y)| \qquad (12.3)$$

Where, $B_R(x, y)$ is the intensity of pixel (x, y) of the background reference frame, and $F_i(x, y)$ is the intensity of pixel (x, y) in the current frame and $D(x, y)$ is the difference.

- **Pixel classification:** The difference $D(x, y)$ is compared with the threshold T for pixel classification in terms of the background and the object frame. The pixel of each current frame has been classified at run-time as.

```
if (D(x,y) ≥ T)
objectFrame (x, y) = 1//Foreground
else
objectFrame (x, y) = 0//Background
end
```

If $D(x, y) \geq T$, the object is classified in the foreground region. If $D(x, y) < T$, the object is in the background region [26, 29].

- **Background maintenance:** This section focuses on computing the histogram of B_R reference frame. The histogram of F_i of current frame is computed in background maintenance. Average of both histogram H_{av} is computed as

$$H_{av} = \frac{H_{B_R} + H_{F_i}}{2} \qquad (12.4)$$

Where H_{av} is the average histogram, H_{BR} is the histogram of reference frame, and H_{Fi} is the histogram of the current frame.

i. *Update learning rate* (α): The learning rate $\alpha_{\text{adaptive}}(x, y)$ must be updated, and depends upon two weighted parameters:

$$\alpha_{adaptive}(x, y) = w_1\alpha_1 + w_2\alpha_2 \qquad (12.5)$$

Where w_1 and w_2 are two weight parameters for $\alpha1$, which is dependent on the magnitude of $D(x, y)$ and α_2 respectively, and $w_1 + w_2 \leq 1$.

$$\alpha_{adaptive} = \frac{1}{2}\left[mean\left[\frac{(1 - \alpha_{adaptive})}{H_{av}} + \alpha_{adaptive} \right] \right] \qquad (12.6)$$

The adaptive learning rate is updated after calculating α adaptively with two weighted parameters.

ii. *Update background model:* The background model is updated pixel by pixel in every frame after updating the adaptive learning rate $\alpha_{adaptive}$.

$$B_R = \alpha_{adaptive} f_i + (1 - \alpha_{adaptive})B_R \tag{12.7}$$

The threshold is obtained using average histogram of each frame in the background [6, 7, 26, 29].

iii. *Threshold updating:* The threshold is updated using the following equation.

$$T = k_1 * T$$
$$T = k_2 * T \tag{12.8}$$

Where $k_1 = 0.5$ and $k_2 = 2.5$, k_1 and k_2 are constant variables used for normalization. The concept is inherited from [26, 29].

12.3.3 Post Processing

We have applied enhancement-based techniques for this work. During pixel classification at run-time, some false alarms have been generated. These alarms cause some interior pixels of the object region to be falsely classified as part of the background. So, holes were generated in the interior region. Here, we have applied a flood-fill algorithm to fill these interior holes of the resulting output.

12.3.3.1 Run-Time for Haque et al. [7] Method

In Table 12.1, we used different datasets and performed experimental analysis in terms of the execution time. Here, we have also presented the execution time of existing methods, i.e., Haque et al. [7].

12.3.3.2 Run-Time for the Proposed Method

In Table 12.2, we used different datasets and performed experimental analysis in terms of the execution time. Here, we have also presented the execution time of one of the proposed methods.

TABLE 12.1

Time Analysis for Haque et al. [7]

Dataset	Number of Frames for BGM	Total Frames in Dataset	Total Time (seconds)	Time per Frame (sec/frame)
Time of day	20	5890	1208.30027	0.2051
Bootstrap	20	3055	854.131022	0.2796
Light switch	20	2715	741.570218	0.2731
Foreground aperture	20	2113	735.238463	0.3480
Moved object	20	1745	559.669293	0.3207
Camouflage	20	353	260.339365	0.7375

TABLE 12.2

Time Analysis for the Proposed Work

Dataset	Numbers of Frames for BGM	Total Frame in Dataset	Total Time (seconds)	Time per Frame (sec/frame)
Water surface	20	1060	10.8785	0.01026
Fountain	20	723	9.0878	0.01257
Light switch	20	2915	35.2101	0.01208
Foreground aperture	20	2313	34.7935	0.01504
Camouflage	20	553	06.5034	0.01176

Table 12.1 and Table 12.2 represent the run-time analysis for Haque et al. [7] and proposed work. Both tables also focus on the following parameters.

- Total number of frames used for background modeling.
- Total number of frames available in the dataset.
- Total time taken to run each dataset.
- Time per frame by each method over considered dataset.

All the above parameters have been explored in Table 12.1 and Table 12.2. The proposed method is experiments using gray color instead of colored frames. So, it takes less time for execution. Here, Table 12.2 clearly indicates the minimum running time taken by the proposed method. Thus, the run-time performance of the proposed work is better as compared to other existing methods. In other words, the proposed method demonstrates superior performance in terms of run-time.

12.4 Experimental Setup and Results Analysis

In this section, we evaluate our method and compare it with the other algorithms, such as Haque et al. [7] and Stauffer et al. [6]. We conducted this experiment using some video sequences. We have taken five datasets, including three from Microsoft-Wallflower [24] and two from I2R dataset [25] as mentioned in Table 12.2. The considered datasets contain the intensity value of each pixel in the range [0, 255]. Each frame was represented in the gray-scale format.

12.4.1 Requirement

This section includes the detailed introduction of the software and hardware platforms. All the experimental analysis was implemented on MATLAB®-2014a software and the Windows 8 operating system environment that runs with 2.33 GHz speed and 8 GB RAM over an Intel *(R)* core-2 duo processor. The experimental results are evaluated qualitatively and quantitatively. The experiment is performed over problematic frame sequences that are taken from well-known datasets. These are publicly available for noncommercial purposes. The considered frame sequences are provided by the following datasets.

- Microsoft's Wallflower dataset
- I2R dataset

12.4.2 Qualitative Analysis

In qualitative analysis, the evaluation results are shown in Figure 12.2 for all five datasets at the learning rate, $\alpha = 0.01$. The experimental results perform against the other considered methods in terms of error measurement. The error rate, i.e., the percentage of pixels detecting false-positive alarms as foreground, and background as false-negative alarms, and their average is reported for all datasets. The proposed technique evaluated the best results for error rate among four out of five datasets.

12.4.3 Quantitative Analysis

This section measures the performance analysis for the following parameters: *FP_Error* is the number of false-positive errors, *FN_Error* is the number of false-negative errors, *TPR* is the true-positive rate, *FPR* is the false-positive rate, respectively. The following indexes are used for total error analysis. To evaluate the precision, recall and F-measure were done with these parameters. Precision is the average value of pixels, whereas recall is the average value of pixels. *F-measure* has been calculated with the harmonic mean of precision and recall. The maximum average value has been compared with the state-of-the-art method that is shown in Figure 12.3 (precision-recall curve). Another performance metric is also evaluated for foreground detection. The average value of percentage of bad classification (PBC) is considered in the video sequence.

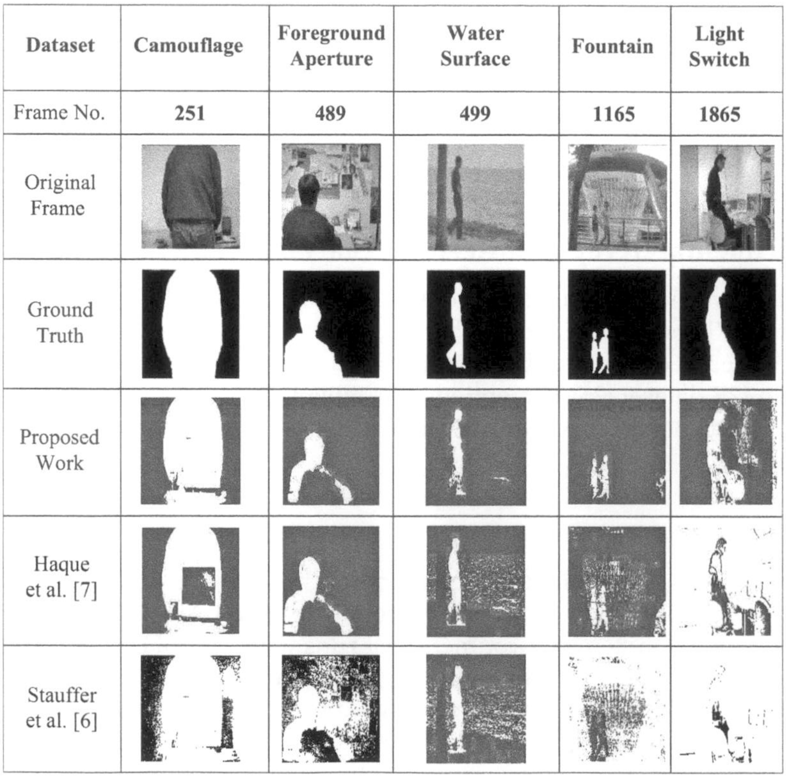

FIGURE 12.2
Performance of Qualitative Analysis: Original Frame; Ground Truth; Proposed Work; Haque et al. [7], Stauffer et al. [6].

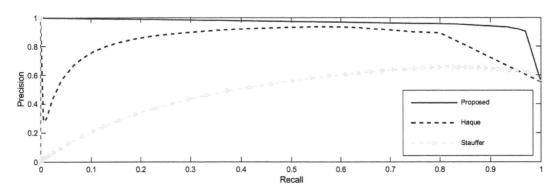

FIGURE 12.3
Precision-recall curve for existing methods.

In the ROC curve (Figure 12.4), the x-axis represents the false-positive rate (FPR) in which a proportion of background pixels is classified as foreground, whereas the y-axis represents the true-positive rate (TPR) in which a proportion of foreground pixels is classified as background.

In the ROC curve, the area under the curve applied for evaluating the performance where accuracy > 0.95 shows better prediction in real-time applications.

12.4.4 Observations

The quantitative observations of the proposed work are described in Table 12.3 and Table 12.4. The varying threshold solves the problem of constant threshold, which has been applied in several methods. The evaluated results of the proposed work are better in terms of error, precision, recall, and F-measure, as shown in Figure 12.2. The shape information is more effective and clear than the other methods.

The F-measure is shown in Table 12.4 for the proposed work, which has a maximum value indicating the classified pixel values that are more accurate and informative. Similarly, accuracy has the maximum value for the proposed method. According to the ROC curve shown in Figure 12.4, the proposed system presents the maximum value. So, the performance of the proposed work is effective and good. The benefits of this work are easy to implement,

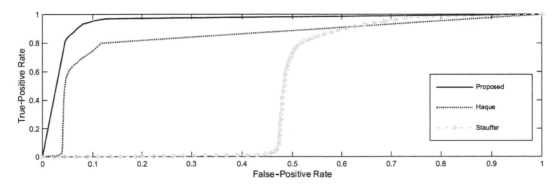

FIGURE 12.4
ROC curve for existing methods.

TABLE 12.3

Error Analysis for Proposed Work, Haque et al. [7] and Stauffer et al. [6]

Methods	FP_Error Average	FN_Error Average	Total_Error Average
Proposed	2.00606	2.8092	4.81524
Haque et al. [7]	16.3492	5.50454	23.05376
Stauffer et al. [6]	39.3912	2.4571	41.84832

TABLE 12.4

Average Performance of Proposed Work, Haque et al. [7], and Stauffer et al. [6]

Methods	Precision Average	Recall Average	F_Measure Average
Proposed	0.8298	0.81328	0.80732
Haque et al. [7]	0.51836	0.66204	0.50976
Stauffer et al. [6]	0.30998	0.72972	0.4074

TABLE 12.5

Performance Analysis for Proposed Work and Considered Methods

Methods	PBC Average	Accuracy Average
Proposed	5.56206	0.94438
Haque et al. [7]	25.89616	0.74104
Stauffer et al. [6]	51.59922	0.48402

simple, and faster based on the gray-scale format. Comparisons of the evaluated results of the proposed work and all considered methods are available in Tables 12.3–12.5.

The Stauffer et al. [6] method developed the adaptive background subtraction using GMM that detects some minor movement because of fluctuations in the background and dynamic background. Hence, false-positive alarms are generated. The Haque et al. [7] method improves the method of Stauffer et al. [6]. Hence, false-positive alarms are still generated. The proposed method improves the work of Stauffer et al. [6] and Haque et al. [7] and generates good results. Hence, there are fewer false-positive alarms, which minimizes execution time and speeds up the performance.

12.5 Merits and Demerits

The experimental study also reveals the merits and demerits of this work as demonstrated below.

1. This work is executed on gray-scale pixels in which each pixel of every frame was executed in iteration form that reduces the running time and speeds up the performance.

2. At execution time, there is no need to consume extra space in the proposed work as compared to the Stauffer et al. [6] and Haque et al. [7] methods. Thus, the proposed work takes less running time for execution.

3. In literature and analysis of practical experiments, it has been observed that the results of Stauffer et al. [6] and Haque et al. [7] are based on GMM, which is a very effective method. The Gaussian has some limitations that deal with dynamic backgrounds.

4. During the testing phase, the varying threshold is used to resolve the problems of fixed and constant threshold at execution time.

5. The proposed method gives more accurate results than other methods for the applications. The experimental results of the proposed work have minimum errors, and cover maximum values over the ROC curve. Therefore, according to the experimental analysis of the measurement, the overall performance of this work is better than the other existing scheme.

12.6 Limitations

Our method works for video with a moving camera but due to unavailability of hardware resources and thermal cameras, we have not analyzed with a moving camera. This method has not worked well for smaller moving objects and objects that are far away from the camera.

12.7 Conclusion and Future Work

The background subtraction technique is generally used for moving object detection with a static camera. In case of dynamic backgrounds, the proposed work handles the noise and variations of the background. This proposed method resolved the problem of fixed threshold and learning rate by providing an automatic updating scheme, i.e., adaptive scheme. According to the experimental analysis, the proposed work is highly perceptual, robust, and generates more accurate results compared with the considered peer methods.

In the future, this work can be developed on a GUI-based framework for moving object detection using IoT devices. Deep learning-based techniques can be attempted for enhancement of results of the work.

Acknowledgments

The author is very grateful to Microsoft's Wallflower dataset (Toyama et al., 1999) team for sharing and allowing a public dataset along with ground truth. The authors also express their gratitude to the change detection dataset (Goyett et al., 2012) community for providing support for analysis of this work.

References

1. Javed, S., Mahmood, A., Maadeed, S. A., Bouwmans, T., & Jung, S. K. (2019). Moving object detection in complex scene using spatiotemporal structured-sparse RPCA. *IEEE Transactions on Image Processing, 28*(2), 1007–1022.
2. Haritaoglu, I., Harwood, D., & Davis, L. S. (2000). W4: Real-time surveillance of people and their activities. *IEEE Transactions on Pattern Analysis and Machine Intelligence, 22*(8), 809–830.
3. Stauffer C., & Grimson, W. E. L. (2000). Learning patterns of activity using real-time tracking. *IEEE Transactions on Pattern Analysis and Machine Intelligence, 22*(8), 747–757.
4. Jiajia, G., Junping, W., Ruixue, B., Zhang, Y., & Yong, L. (2017). A new moving object detection method based on frame-difference and background subtraction. *IOP Conf. Series: Materials Science and Engineering, 242,* 1–4.
5. Sandeep, S. S., & Susanta, M. (2017). Foreground detection via background subtraction and improved three-frame differencing. *Arabian Journal for Science and Engineering, 82*(8), 3621–3633.
6. Stauffer, C., & Grimson, W. E. L. (1999). Adaptive background mixture models for real-time tracking. *IEEE Computer Society Conf. on Computer Vision and Pattern Recognition, 2,* 246–252.
7. Haque, M., Murshed, M., & Paul, M. (2008). On stable dynamic background generation technique using Gaussian mixture models for robust object detection. In 5th Int. Conference on Advanced Video and Signal Based Surveillance, IEEE, pp. 41–48.
8. Jung, C. R. (2009). Efficient background subtraction and shadow removal for monochromatic video sequences. *IEEE Transactions on Multimedia, 11*(3), 571–577.
9. Hu, W., Tan, T., Wang, L., & Maybank, S. (2004). A survey on visual surveillance of object motion and behaviour. *IEEE Transactions on Systems, Man, and Cybernetics, 34*(3), 334–352.
10. Lin, H., Liu, H., & Chuang, J. (2009). Learning a scene background model via classification. *IEEE Transactions on Signal Processing, 57*(5), 1641–1654.
11. Ming, T., & Chih, L. (2009). Independent component analysis-based background subtraction for indoor surveillance. *IEEE Transactions on Image Processing, 18*(1), 158–167.
12. McHugh, J. (2008). *Probabilistic methods for adaptive background subtraction* (Master thesis). Boston University.
13. Shimada, A., & Arita. D. (2006). Dynamic control of adaptive mixture-of-Gaussians background model. In IEEE International Conference on Video and Signal Based Surveillance (AVSS'06), Piscataway, pp. 1–5.
14. Lee, D. S. (2005). Effective Gaussian mixture learning for video background subtraction. *IEEE Transactions on Pattern Analysis and Machine Intelligence, 27*(5), 827–835.
15. Wang J., Bebis, G., & Miller, R. (2006). Robust video-based surveillance by integrating target detection with tracking. In IEEE Computer Vision and Pattern Recognition Workshop CVPRW, Piscataway, p. 137.
16. Ridder, C., Munkelt, O., & Kirchner, H. Adaptive background estimation and foreground detection using Kalman-filtering. In International Conference Recent Advances in Mechatronics, ICRAM 95, pp. 193–199.
17. Das, B. P., Jenamani, P., & Mohanty, S. K. (2017). On the development of moving object detection from traditional to fuzzy based techniques. In 2nd International Conference on Recent Trends in Electronics, Information & Communication Technology, IEEE, pp. 658–661.
18. Wang, L., Hu, W., & Tan, T. (2003). Recent developments in human motion analysis. *Pattern Recognition, 36*(3), 585–601.
19. Wren, C., Azarbayejani, A., Darrell, T., & Pentland, A. (1997). Pfinder: Real-time tracking of the human body. *IEEE Transactions on Pattern Analysis and Machine Intelligence, 19*(7), 780–785.
20. Lo, B. P. L., & Velastin, S. A. (2001). Automatic congestion detection system for underground platforms. In Intelligent Multimedia, Video and Speech Processing, Proceedings of IEEE 2001 International Symposium, pp. 158–161.

21. Glocker, B., Komodakis, N., Tziritas, G., Navab, N., & Paragios, N. (2008). Dense image registration through MRFs and efficient linear programming (PDF). *Medical Image Analysis Journal, 12*(6), 1–24.
22. Gao, G., Wang, X., & Lai, T. (2015). Detection of moving ships based on a combination of magnitude and phase in along-track interferometric SAR-part I: SIMP metric and its performance. *IEEE Transactions on Geoscience and Remote Sensing, 53*(7), 3565–3581.
23. Haines, T. S. F., & Xiang, T. (2014). Background subtraction with Dirichlet process mixture model. *IEEE Transactions on Pattern Analysis and Machine Intelligence, 36*(4), 670–683.
24. Toyama, K., Krumm, J., Brumitt, B., & Meyers, B. (1999). Wallflower: Principles and practice of background maintenance. *7th International Conference on Computer Vision, September, 1*, 255–261.
25. I2R Dataset. http://perception.i2r.a-star.edu.sg/bk_model/bk_index.htm
26. Yadav, D. K., & Singh, K. (2015). A combined approach of Kullback-Leibler divergence method and background subtraction for moving object detection in thermal video. *Infrared Physics and Technology, 76*(1), 21–31.
27. Ng, K. K., & Delp, E. J. (2010). Object tracking initialization using automatic moving object detection. *Proceedings of SPIE/IS&T Conference on Visual information Processing and Communication, 7543*, San Jose, CA.
28. Lee, S., & Lee, C. (2014). Low complexity background subtraction based on spatial similarity. *EURASIP Journal of Image Video Processing*, 2–16.
29. Zhou, X., Yang, C., & Yu, W. (2013b). Moving object detection by detecting contiguous outliers in the low-rank representation. *IEEE Transactions on Pattern Analysis and Machine Intelligence, 35*(3), 597–610.
30. Comaniciu, I., Dorin, L., & Meer, C. (2002). Mean shift: A robust approach toward feature space analysis. *Pattern Analysis and Machine Intelligence, IEEE Transactions, 24*(5), 603–619.
31. Yadav, D. K., Sharma, L., & Bharti, S. K. (2014). Moving object detection in real-time visual surveillance using background subtraction technique. In 14th International Conference on Hybrid Intelligent Systems, IEEE, Kuwait, pp. 79–84.
32. Yadav, D. K. (2014). Efficient method for moving object detection in cluttered background using Gaussian mixture model. In 3rd International Conference on Advances in Computing, Communications and Informatics, IEEE Man Systems and Cybernetics, India, pp. 943–948.

13

Hadoop Framework: Big Data Management Platform for Internet of Things

Pallavi H. Bhimte[1], Pallavi Goel[1], Dileep Kumar Yadav[1], and Dharmendra Kumar[2]

[1]*Dept. of CSE, Galgotias University, Greater Noida, India*

[2]*Delhi Technical Campus, Greater Noida, India*

CONTENTS

13.1 Introduction

Big data is really a whole bunch of data collection, and Hadoop is a framework that deals with the huge amount of data. Big data analytics includes the effective design algorithms that integrate the data and solve the problem of hidden values [1]. Big data focuses on learning, knowledge discovery, data modeling, data transformation, prediction, and visualization of the data [1, 2]. This huge amount of data can be related to social media profiles, their connections, their messages, or their daily posts, along with shares, likes, and comments. Big data includes diverse categories of datasets in the power grid; a variety of questions being asked on Bing, DuckDuckGo, Ecosia, Exalead, Quora, Yahoo!, and Google search engines; and storing and accessing data on most popular Internet cloud-based platforms such as Google Cloud SQL, Amazon, EnterpriseDB, Garantia Data, and many more. It also has numerous opportunities, challenges, and growing research areas [2–5]. Big data is involved with stock buyers associated with stock sellers in stock exchange market data, and with transport-related data that keep records of vehicles' type, model, travel distance, and capacity, along with other factors.

Big data includes the data generated by various kinds of devices and applications. Some areas that come under the umbrella of big data, are discussed below [1–3, 6–9].

- **Black box data:** Such data come from various component of airplanes, helicopters, and jets.

- **Social media data:** Numerous sources of social media, for instance, Facebook, Twitter, WhatsApp, Digg, Viber, Quora, Flickr, Tumblr, Instagram, YouTube, LinkedIn, and Periscope, are responsible for holding the data accumulated in the form of posts uploaded by people along with the number of likes, reactions, comments, shares, and views.

- **Stock exchange data:** Here, stock exchange data depict information related to buying and selling decisions made by customers based on shares of various organizations.

- **Power grid data:** It holds the information of base stations, which is consumed by a particular node.

- **Surveillance data:** It represents the data captured though various cameras or sensory devices, such as closed-circuit television (CCTV), Internet protocol, and wireless, which is used for surveillance of some incident based on past records. Such data may belong to video surveillance applications, such as transportation, indoor-outdoor, border, seaport, restricted zones, deep zones, hilly regions, crowded places, and navigation [10, 11].

- **Transport data:** Transport data may include the information related to the model, geographical local, color, capacity, speed, distance, and availability of vehicles [10, 11].

- **Search engine data:** A search engine searches and retrieves lots of data (structured, unstructured, or semistructured) from different databases, which can be classified per requirements of end users [12].

- **Big data Hadoop:** The big data Hadoop may consist of various technologies, such as Hadoop 1.x, Hadoop 2.x, Hive, Pig, Flume, Sqoop, Spark, Impala, Mahout, Kafka, HDP 2.0, CDH 5, and so on [6, 13–16].

- **Data integration and visualization:** Software industries use Informatica, Tableau, and Pentaho tools for integration of data and their visualization.

- **NOSQL and SQL databases:** Database-related queries are handled by using MongoDB, MySQL, and PostgreSQL [1–3, 15, 17].

- **Programming:** Numerous languages support big data; some languages are PHP, Java, and Python.

Data that are to be stored, organized, or managed can be unstructured, semistructured, or structured. Various techniques help in analyzing the data, categorizing them, and storing the data in a systematic manner. Through Hadoop, big data can easily be handled. Hadoop is an open-source program under the Apache license, including MapReduce and hadoop distributed file system (HDFS). Hadoop is a framework that finds solution for problems of massive amounts of data and computation through a network system facilitation. Hadoop can be used to manage the three V's of data—volume, variety, and velocity—where volume represents the amount of data, variety represents the number of types of data, and velocity represents the speed of data processing. The three core components of the Hadoop framework are HDFS, MapReduce, and YARN; subcomponents are: Hive, HBase, Flume, Oozie, Ambari, Pig, Avro, Mahout, Sqoop, HCatalog, and BigTop for processing, analyzing, and storing large volumes of data.

The Hadoop system solves the big data problems related to volume, variety, velocity, and values. Hadoop has some basic challenges, as mentioned below [1–3, 6, 7, 15, 17].

- Hadoop is a very complex distributed system and consists of low-level APIs.
- The end-to-end solutions for automated testing are impossible or impractical.
- Skilled technocrats are required for Hadoop.
- Hadoop is a diverse collection of various open-source projects to know and understand the hand-coding integration and multiple technologies between them.
- Business logic and infrastructure-related APIs have no clear separation and burdening.
- A significant effort is wasted on basic jobs, such as extract, transport, load (ETL) and data ingestions.
- Various processing paradigms need data to be stored in some specific manner.
- Hadoop is an offline storage along with batch analytics.
- Understanding of transferring from proof-of-concept to production is a difficult task that can take a long time.
- Most data patterns are common, but they do not support data correctness and consistency.
- Real-time and batch ingestion are needed for deeply integrating of several components.

To address the above challenges, software technocrats at the enterprise level need to go with specific tools [7, 12, 13, 15, 17]. Some commercial Hadoop software tools are available for handling big data issues, such as Cask, Mica, Bedrock, Talend, hTrunk, Pentaho, and Informatica big data management. These may provide the true benefits of Hadoop's power.

13.2 IoT Devices and Data

In real-time systems, IoT devices generate huge amounts of data in terms of volume, variety, velocity, variability, veracity, and other inherent characteristics of data. These IoT devices are connected devices and the data generated from them have various character-istics, as given below.

- Intermittent data having massive volumes.
- Data generated in terms of streams or batches.
- Varied data sources, such as data generated by numerous sensors in flight.
- Time-series data generated predominantly.
- Diverse nature of data structures and schemas applied.
- Some data may be perishable as the value of data reduces over time.

According to the characteristics of data streams, modern organizations are increasingly moving towards Apache Hadoop and Cloudera. Hadoop and Cloudera are used as stan-dard platforms for data management, such as storage, management, processing, and ana-lyzing the data. The architecture of the Hadoop framework works as a data management platform for IoT.

Various IoT big data cloud providers help in availing the private cloud for IoT data stor-age, and also provide big data processing software, such as "stack" that supports data analytics as well as processing of intelligence over such data. Reasons for using Hadoop in the world of IoT data as a platform for storage, management, and analytics include:

- **Data ingest:** The Hadoop framework easily ingests data generated from multiple sources. It also supports real-time and batch data that ingest from various sensors using Hadoop components, such as Apache Flume and Apache Kafka.
- **Handles variety of data:** It effectively handles numerous IoT data-types along with data structures as well as schemas from intermittent sensors. This type of data consists of pressure, humidity, and temperature to real-time geographical locations or live video streams.
- **Deployment of data:** As per requirements, a hybrid environment is provided to deploy the platform for such data.
- **Serving and insights in real-time:** It also supports real-time based data process-ing and analytics. Such real-time data streaming is processed through Apache Spark (as Spark streaming), while the storage is supported by Apache HBase and Apache Kudu.
- **Secure:** For IoT and Cloudera, the Hadoop platform supports multiple layers of security as well as industry-oriented tools to enable security.
- **Flexible and scalable processing of data:** Based on growth of data, it scales data effectively and enables an enterprise to store huge volumes of data. This platform also allows the combining of IoT/sensor data to execute deeper business insights.
- **Fast analytics:** IoT data are supported and processed by the Hadoop system for business intelligence. The analytics is performed with tools such as Apache Impala.

Huge amounts of IoT data (i.e., big data) are handled and processed through the Hadoop environment. The Hadoop platform is configured in single-node as well as multinode cluster environments. In forthcoming sections, the chapter explores the configuration of the Hadoop platform in both environments.

13.3 Core Components of Hadoop

The Hadoop system has numerous core components that are helpful for performing different tasks. The forthcoming sections explore the specialty features of each component.

13.3.1 Hadoop Distributed File System

The Hadoop distributed file system (HDFS) is used for storing massive amounts of data [12, 16–19]. HDFS can be built on commodity hardware. It has a master-slave architecture, which contains only one master and an infinite number of slaves. The master is named as name node, while various slaves are known as data nodes. Another component, known as the secondary name node, is responsible for taking screen shots of the original name node. The secondary name node stores data that takes checkpoints of the file system metadata present on name node. The biggest cluster found until now is Yahoo!. If the active name node fails, the passive name node comes into existence. Name nodes handle the metadata while data nodes handle the actual data.

The normal window block size is 1KB. In HDFS, block size is 64MB. Therefore, if a file of 200MB is to be stored into HDFS, it will be stored in four partitions, i.e., three 64MB blocks and one 8MB block. The default replication factor in HDFS is 3, i.e., if there are three data nodes namely, datanode1, datanode2, and datanode3, and four blocks are stored under datanode1. Those four blocks will automatically be replicated into the remaining two data nodes. Heartbeat signals are sent by data node to other data nodes every 5 seconds. In case any one of the data nodes fails to send the heartbeat signal, that will mean it is no longer working. Then the remaining other data nodes will come to be aware of this, after which the data nodes containing the replicated blocks of original data node will come into play and the process will continue. The data node will then send block report to namenode of the tasks being performed in datanode. The main function [16, 18–20] of datanode is to create a block, move a block to its required destination, replicate the data of other data nodes, and accept requests from name node. The data node also provides read and write access. The namenode regulates the access to clients and all the operations are done by namenode. The working architecture of HDFS [21] is shown in Figure 13.1.

The process of replication of a particular block from one data node to another is known as pipeline process. Subdirectories are created and logs are generated in Edit log. FSImage and Edit log are stored in local file systems. FSImage files are maintained in name space. The checkpointing is a process in which when the namenode starts, the edit log deletes data after FSImage has received that data. A safe mode state is a state when data have not been replicated yet, and name node keeps the record of data nodes and its blocks. Whenever the resource leaves the threshold state, it enters the safe mode state. If there are three racks, Rack1, Rack2, and Rack3, then Rack2 and Rack3 contain the replicated data nodes of Rack1. A whole rack cannot collapse; only a data node of any of the racks can collapse, but the entire rack cannot collapse.

The trash has the files that were deleted and these files are kept in the trash for around 24 hours and then deleted permanently. When a name node is replicated to another name node that point is known as checksum. The network bandwidth has to be very high because it has to store the large datasets. Because moving computations costs less than moving data, HDFS is best known for storing huge amounts of data. In case of a hardware failure, HDFS detects it automatically and gets rid of it.

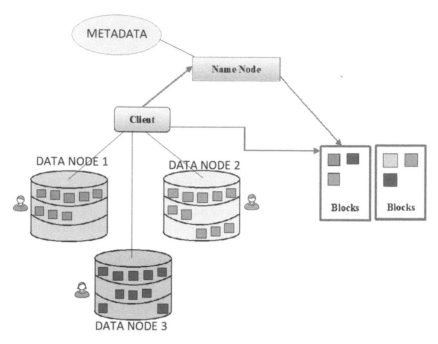

FIGURE 13.1
Basic HDFS architecture.

13.3.2 MapReduce

Based on Java, the MapReduce model facilitates many computers for accessibility. It comprises two basic tasks, i.e., Map and Reduce. Map creates key-value pairs of the data that have been converted to another set of data. Every single element of the data is broken into tuples. These tuples are then combined with the output of Map and an even smaller set of tuples is formed. MapReduce is extensible for a large number of servers in a Hadoop cluster. It is easy to scale data processes over multiple nodes available. MapReduce is also responsible for handling parallel processing of many datasets, which might be distributed across a variety of computers. The basic working of data processing through MapReduce can be understood through Figure 13.2.

MapReduce also has master-slave architecture. The role of master in a MapReduce job is done by Job Tracker and the role of slave is fulfilled by Task Tracker. Job Tracker is responsible for assigning various jobs to different Task Trackers, and also maintains a proper record of jobs assigned to Task Tracker. Task Tracker performs the given task and sends the report to Job Tracker. A job that is to be executed is a program across a dataset of a Mapper and a Reducer. Once a file has been located, the InputSplit function processes the smaller pieces of files for further processing. After the formation of smaller pieces of file, Map and Reduce work together in order to process the data. Map is a process in which files are read and their content is known as well as arranged in a way so that it can be shuffled easily. Shuffling is a process in which same content is grouped together and similar data are separated from the rest of the data. In Reduce, identical data shuffled earlier are collected to form a single group.

MapReduce can handle unstructured data. Map helps in converting the unstructured data to structured data. Every single key-value pair is taken by mapper as input and any number

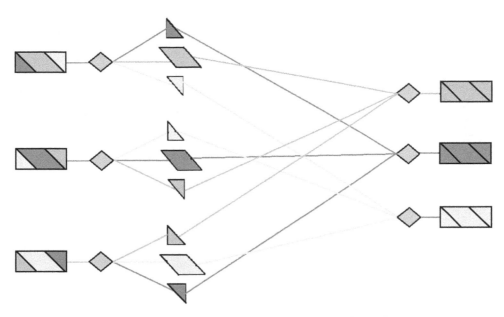

FIGURE 13.2
Basic working environment of MapReduce.

of key-value pairs is produced as output. There can be multiple Task Trackers of a single Job Tracker. For every job, there is one Job Tracker that resides on name node, and there are multiple Task Trackers that reside on data nodes. Multiple tasks run on multiple data nodes. The activities of scheduling tasks to run on different data nodes are managed by Job Tracker.

Task Tracker manages the execution of individual processes residing on data nodes. It also sends heartbeat signals to Job Tracker to make sure that tasks are being performed by Task Tracker. As soon as there is no heartbeat signal received by Job Tracker from a Task Tracker, another Task Tracker that contains the replicated data comes into play. The tasks can be rescheduled on another Task Tracker if one Task Tracker fails to send the heartbeat signal to Job Tracker. It is the responsibility of Task Tracker to send continuous reports to Job Tracker to ensure that tasks are performed. All Task Trackers have to perform Map Task and Reduce Task and send the output to Job Tracker. As soon as the status is updated, Job Tracker receives it and confirms that tasks are being performed on time.

13.3.3 Yet Another Resource Negotiator

Yet another resource negotiator (YARN) forms groups, known as clusters, for storing a variety of data as well as properly managing it. Resources can be easily managed with help of YARN. It provides a platform to perform operations consistently. Its working is shown in Figure 13.3. Data are governed with help of data governance tools provided by Hadoop clusters. Data remain safe as they also provide proper security. YARN helps in dynamic allocation of cluster resources. It allows graph processing, stream processing, and batch processing to run and store data in HDFS. Apache YARN is a data operating system of Hadoop 2X. YARN architecture consists of a Resource Manager and a Node Manager. YARN also has a master-slave architecture in which one Resource Manager can have many Node Managers. A Resource Manager consists of Scheduler and an Application Manager. A Scheduler allocates the resources to running applications. An Application Manager is

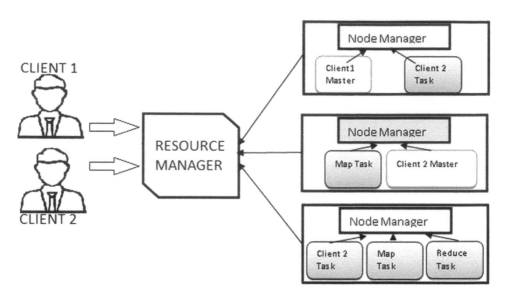

FIGURE 13.3
Working framework for YARN.

responsible for running application masters on the cluster and keeps a check on the work. An Application Manager also handles the situation where there is a chance of failure, and restarts Application Master for continuation of the processes. Node Manager is the slave daemon of YARN. Node Manager monitors resource usage and reports the status to Resource Manager. Node Manager also keeps a check on the health of the node. An application consists of one or more tasks that are created by the client. In a YARN cluster, Application Master is responsible for running all kinds of applications, which in turn helps to coordinate all sorts of tasks that are to be performed. The Application Master works with the Node Manager on specific framework to get resources from the Resource Manager and also to manage the various task components.

If a failure is detected, Non-Work-Preserving Resource Manager restart and Work-Preserving Resource Manager restart are the two kind of restarts. In a Non-Work-Preserving Resource Manager restart, the same information from the state store will be reloaded on previously running apps. In Work-Preserving Resource Manager restart, container request from Application Masters and container status from Node Manager constitutes the running state. YARN has a higher degree of compatibility, better cluster utilization, scalability, and multitenancy.

13.4 Subcomponents of Hadoop

This section depicts the various elements of Hadoop, such as Hive, HBase, Pig, Flume, Sqoop, Oozie, and Ambari, along with their features.

13.4.1 Hive

The Hive is a data warehouse infrastructure tool where static data exist. Structured data can easily be processed into Hive. Concepts of core Java, SQL, and any Linux operating system are required for working with Hive. In Hive, a database can be created and dropped (deleted) as well as a table can be created, altered, and dropped. Various data types, operators, and functions are used in it. The query language used in Hive is known as HiveQL.

The content of a table can be directly downloaded into a local directory or as a result of queries to a HDFS directory. Evaluation of functions can take place in it as well as various arithmetic, logical, and logical operations can be performed on tables. Two data tables can be joined as well as results of one query can be stored into another table. External tables can be created that point to a specific location in HDFS. There are three different formats in Hive:

1. **Tables:** It contains rows and columns similar to RDBMS. These rows and columns store the details of the file system. Data can be stored in the form of tables in Hive.
2. **Buckets:** The data stored in Hive can be divided into buckets, which store files in partitions in the suppressed file system.
3. **Partitions:** Hive data tables can have partitions. Subdirectories can be created and file system data can be arranged in the form of rows and column in tables of Hive, which can further be partitioned more than once.

13.4.2 HBase

HBase is an open-source nonrelational database that gives real-time read/write access to many large datasets. HBase is Java-based not only SQL-based. HBase allows for dynamic changes, and can be utilized for standalone applications. HBase tables containing rows and columns are labeled. Each intersection of a row and column is versioned. This version is auto-signed and time stamped by default. The row keys are byte arrays and column keys are grouped into families. Each family has a specific name having certain information, like name, age, date of birth, and so on [22]. The data that are stored into the table are stored horizontally. All data are stored in a single table. For every single region, there exists a MemStore. Without any joining, families are created for similar kinds of data stored in the tables of HBase. In columns, versions can be created for updating data; by default it is three. For every version, there is a unique timestamp. Using row keys, any cell in the column family can be accessed. HBase is fault-tolerant, fast, and usable. Replication is possible across the data center, so lost data can be regained.

A table once created in HBase can be listed, disabled, enabled, described, altered, and dropped. Data that are stored in the table can be created, updated, read, deleted, scanned, truncated, and stay secure throughout the storing process as well as after the data are stored. There are various built-in classes, constructors, and methods in HBase for storing and processing of data. HBase is ideally suited for random read and write of data stored in HDFS. It is dependent on zookeeper and the authority of the cluster state is also managed by zookeeper instance. A four-dimensional data model of HBase contains:

1. **Row key:** Every row has its own row key, and is treated as a byte array.
2. **Column family:** Every column has a column name automatically assigned to it.
3. **Column qualifier:** Column families define columns, which are called column qualifiers.
4. **Version:** Every column can have one or more than one version, and specific data can be accessed by knowing the version of a column qualifier.

13.4.3 Pig

Pig is a procedural data flow with high-level scripting language mainly used for programming with Apache Hadoop. It can be used by those who are not familiar with Java and are familiar with SQL scripting language, also known as Pig Latin. There is a user-defined function (UDF) facility in Pig that invokes code in many languages, like JRuby, Jython, and Java. Pig operates on the client side of a cluster and supports the Avro file format. It can easily handle large amounts of data and can be used for ETL data pipeline; research on raw data can be processed again and again. Data in Pig Latin can be loaded, stored, streamed, filtered, grouped, joined, combined, split, and sorted. A Pig program can be run in three different ways:

1. **Script:** A file containing Pig Latin commands.
2. **Grunt:** As a command interpreter, it executes the command.
3. **Embedded:** A pig program can also be executed as part of a Java program. Pig is used in Dataium to sort and prepare data, in "People you may know" in LinkedIn, and in PayPal to analyze transactional data and attempt to prevent fraud.

13.4.4 Sqoop

The transfer of large data from Apache Hadoop to structured datastores is done by Sqoop, such as a relational database on enterprise data warehouses. It can also extract data from Apache Hadoop and export the data into relational databases [23]. Sqoop even works along with SQL databases, for instance, Teradata, Netezza, Oracle, MySQL, PostgreSQL, and HSQLDB. It helps in the process of ETL. In this process, Sqoop basically extracts the data from relational datastores, transforms them in a way that can be efficiently used by Hadoop, and then loads the data into HDFS. It provides a command life interface, and Sqoop commands written by an interpreter are executed one at a time.

Sqoop helps to import databases and tables into HDFS. It can import the minor data as well as data in bulk. It imports and maps SQL directly into Hive and HBase. Java classes are generated as well for interaction with data programs. DAO classes are generated automatically by Sqoop for getter and setter methods to initialize objects. This can be achieved by the codegen tool available in Sqoop for checking if Java lost the Java code, which will further create new versions of Java with default more than one character field in between various fields. Also using the eval in Sqoop, SQL queries can easily be evaluated for DDL or DML statements.

13.4.5 Flume

Flume is a service for running logs into Hadoop. It is a tool that helps in transferring the streaming data. Flume helps in collecting the files to a single location. It is used to move data from one location to another. There is a data source from where the data is to be collected, and every data unit is considered an event. When multiple events are collected from a specific data source, these can be transferred to a single location with the help of Flume. Flume consumes the events and transfers them to the destination in order. It helps in real-time data transfer, which prevents any kind of data loss. Flume is made up of three components:

1. **Source:** Flume collects data from the source.
2. **Channel:** Temporary storage for different events that take place.
3. **Sink:** It retrieves the events from a channel and passes them on to the next agent.

13.4.6 Oozie

Oozie [24, 25] is a web application based on Java that schedules Apache Hadoop jobs. Multiple jobs are combined in specific order into a single logical unit of work. It works independent of jobs. Oozie and the Hadoop stack combine to support various types of Hadoop jobs in a creative manner (for example, streaming MapReduce, Java MapReduce, Hive, Pig, Distcp, and Sqoop) as well as jobs related to a system (for example, Java programs and shell scripts). It is extensible and reliable.

Users can make directed acrylic graphs (DAGs) of workflows, which can further be run in Hadoop where there are sequences of actions in specific arrangements according to the tasks that are to be performed [24, 25]. Jobs on Oozie can be started, stopped, suspended, and rerunned. The workflow defined by the client can run the jobs on Oozie. Even the clusters formed help in defining a specific job in Oozie. These jobs can be processed either early or later on. There are two types of nodes in a graph:

1. **Control node:** It provides rules for the beginning and ending of a workflow.
2. **Action node:** It is used for execution of tasks.

An Oozie job is when multistage Hadoop combines within a single job. Two or more jobs can run parallel to each other in Oozie [23]. Three types of jobs in Apache Oozie are:

1. **Oozie workflow Hadoop:** Collection of actions with help of DAG
2. **Coordinator job:** Multiple workflow jobs are managed
3. **Oozie bundle:** Collection of many coordinator jobs

13.4.7 Ambari

Ambari came into existence when it became problematic for the clusters to handle so many nodes and requests simultaneously. It is one of the foremost projects run by Apache Software Foundation. Ambari can create single-node cluster and multiple-node clusters. Single-node cluster is where all the daemons run on the same machine but on different ports. Multiple-node cluster is where two or more clusters are running simultaneously [26]. The Ambari server is responsible for communicating with the agents that are installed on each node on the cluster.

Ambari can run on any platform, i.e., Windows, Mac, and many more. Any Ambari application can be customized. It maintains its versions without any need for external tools, like Git. The system can recover from any kind of failure experienced during the work being performed on it. Recovery is possible based on actions and desired states. Ambari is used extensively by database professionals, Hadoop administrators, Hadoop testing professionals, and mainframe DevOps professionals. Big data innovators, like Hortonworks, are working on Ambari to make it more scalable to support more than 2000 and 3000 nodes seamlessly. Ambari creates clusters of various kinds of data depending on the category into which they fall. Master-slave is assigned to configuring services. Ambari has advanced security constructs, such as Kerberos and Ranger. The status information of various clusters of HDFS, YARN, and HBase is easily visible. The restful APIs provide integration with other operational tools.

13.4.8 Avro

Avro is an open-source project in which source code is modifiable that helps in storing the data in a format where the data are translated for storing purposes. It provides data exchange in Apache Hadoop. A message or file contains data along with a data definition. The exchange of big data between programs can be written in any language. It provides fast, binary data on schema. When data are stored in Avro, the schema will be stored at the same time so that the files can be processed with other programs. Avro accepts schemas as inputs, which are created in JavaScript Object Notation (JSON) document format. JSON defines data types and protocols in Avro. For establishing a communication between Hadoop nodes and Hadoop services, a wire format is responsible for it with a serialization format provided [27]. It also supports Ideal Description Language syntax, known as Avro IDL. Schema is used while deserializing the data.

Avro can easily handle any number of data formats, and it can further be processed in a variety of languages. The process that usually takes place in Avro is the creation of a table. Content or data are loaded into the table, and certain queries are performed. The file is downloaded with the Avro extension, Avro schema can then be viewed and converted into JSON, which can be finally be viewed in JSON format.

13.4.9 Mahout

Mahout is an algorithm library for applying machine learning on top of Apace Hadoop. Machine learning is the concept of a machine that can learn tasks to be performed on its own. It uses the knowledge gained in an efficient way to take the next step that can be predicted by the machine and performed for best outcomes [19, 20]. Through machine learning, specific patterns are analyzed by the machine itself for the required outcome. Data science tools are provided by Mahout for finding these patterns. The accuracy of the pattern depends on the data that you provide to the machine. There are four types of data science use cases in Mahout:

1. **Collaborative filtering:** Product recommendations based on user behavior.
2. **Clustering:** Organizing the data into a group of similar data or related to that data.
3. **Classification:** Classifying the unclassified data into best categories.
4. **Frequent itemset mining:** Analyzing items in a group.

Machine learning is used in various applications, for instance data mining, language processing, games, pattern recognition, forecasting, visual processing, expert systems, robotics, video surveillance, Siri, Cortana, Alexa, and many more. Mahout provides the main algorithms for various types of clustering, filtering, and classification that are implemented on top of Apache Hadoop.

13.4.10 HCatalog

HCatalog is a system that helps in managing tables for a wide Hadoop platform. It is basically a metadata, which stores data in any given format. Structured as well as unstructured data can be stored in HCatalog. It has an interface for external use that enables the current tools to interact with Hadoop. HCatalog has the same interface as that of Hive but abstracts it for data beyond Hive. Hive enables the right tool for the right job, helps in capturing processing states to enable sharing as well as integrating Hadoop with everything. A table of stored contents can be created, altered, viewed, partitioned, indexed, and loaded using HCatalog. HCatalog also supports RCFile, ORCFile CSV, Parquet JSON, and SequenceFile formats. It provides a relational view of stored data. Data are stored in tables, and these tables can further become part of databases. Tables can easily be divided into various categories. HCatalog incorporates Hive's DDL.

13.4.11 BigTop

BigTop is a project for the developing packages and tests of the Hadoop ecosystem. It gathers the core Hadoop components to ensure working configurations. BigTop helps in building a community around packaging, deployment, and integration of projects in the Apache Hadoop system. It focuses on the whole rather than focusing on a single project.

13.5 Single-Node Cluster

A single-node cluster includes all the paramount daemons (NameNode, DataNode, Resource Manager, and Node Manager) run on the same machine but on different ports. This chapter demonstrates the configuration and basic steps for configuration in the following figures. The environment setup is shown in Figure 13.4.

```
Step 1: Install JDK8 and Apache Hadoop 2.7.0.
         Run the following commands after you ssh into your machine
$ cd /home/training
$ sudo systemctl stop firewalld
$ sudo systemctl disable firewalld

< set selinux to disabled mode and restart the machine >

$ sudo yum install wget
$ wget --no-check-certificate --no-cookies --header "Cookie: oraclelicense=accept-securebackup-cookie"
http://download.oracle.com/otn-pub/java/jdk/8u102-b14/jdk-8u102-linux-x64.tar.gz
$ tar –zxf jdk-8u102-linux-x64.tar.gz
$ wget https://www.apache.org/dist/hadoop/core/hadoop-2.7.0/hadoop-2.7.0.tar.gz
$ tar –zxf hadoop-2.7.0.tar.gz
$ sudo mv hadoop-2.7.0 /usr/local/Hadoop
$ sudo mv jdk1.8.0_102/ /usr/local/java

Edit "/home/training/.bashrc" file and define the following variable

export JAVA_HOME=/usr/local/java
export PATH=$PATH: $JAVA_HOME/bin
export HADOOP_HOME=/usr/local/Hadoop
export PATH=$PATH: $HADOOP_HOME/bin:$HADOOP_HOME/sbin

After commands of Step1, run the command $ source /home/training/.bashrc
```

FIGURE 13.4
Installation of JDK1.8 and Apache Hadoop 2.7.0.

```
Step 2:
Create the passwordless-ssh key and give them proper permission

$ ssh-keygen -t rsa –P ""
$ cd /home/training/.ssh
$ cat id_rsa.pub >> authorized_keys
$ chmod 755 authorized_keys
```

FIGURE 13.5
Creating the passwordless-ssh key.

Step 2 demonstrates the creation of a passwordless-ssh key. This key is mainly helpful for granting access authority (Figure 13.5).

Step 3 signifies the configuration of Hadoop by editing the hdfs-site.xml (Figure 13.6), mapred-site.xml, yarn-site.xml (Figure 13.7), and core-site.xml (Figure 13.8). After these files are configured and the following changes have been made, then master and slave files are to be created (Figure 13.9). The only step left after that is to start the cluster after formatting the name node.

In the above steps, the manuscript describes the main key steps involved during setting up a single-node cluster. These steps clearly depict the useful steps and procedure through commands. The primary strength of big data Hadoop is to distribute a task across numerous nodes. One more important key point is that the default replication factor for single-node cluster is 1, whereas for multinode cluster it is 3. In a single-node Hadoop cluster, all the sets of services (Namenode, JobTracker, TaskTracker, DataNode, Secondary NameNode) are running and storing the data on Hadoop, and later processing it. In other words, all Hadoop daemons lie on the single machine. In cases of multinode cluster environments, the DataNode and TaskTracker (NodeManager) are running on a commodity computer, and the remaining services need a powerful server to run. In other words, all Hadoop daemons lie on different machines. A single-node cluster is mainly applied to simulate a full cluster, like environment. So, a single-node cluster does offer some benefits over a multinode cluster, as well as some limitations

```
Step 3:
Configure Hadoop
Now edit the following configuration file of hadoop present under
$HADOOP_HOME/etc/hadoop folder

(A) hdfs-site.xml
<property>
<name>dfs.replication</name>
<value>1</value></property><property>
<name>dfs.datanode.data.dir</name>
<value>/home/training/hadoop/data</value>
</property>

<property>
<name>dfs.namenode.name.dir</name>
<value>/home/training/hadoop/name</value>
</property>
```

FIGURE 13.6
Editing hdfs-site.xml.

```
(B) mapred-site.xml
<property>
<name>mapreduce.framework.name</name>
<value>yarn</value>
</property>

(C) yarn-site.xml
(Replace $(hostname) by the hostname of your machine in the below properties)

<property>
<name>yarn.nodemanager.aux-services</name>
<value>mapreduce_shuffle</value>
</property>
<property>
<name>yarn.resourcemanager.address</name>
<value>$(hostname):8032</value>
</property>
```

FIGURE 13.7
Editing mapred-site.xml and yarn-site.xml.

```
(D) core-site.xml
(Replace $(hostname) by the hostname of your machine in the below properties)

<property>
<name>fs.defaultFS</name>
<value>hdfs://$(hostname):54310</value>
</property>
(E) Create a file by name "masters" under $HADOOP_HOME/etc/hadoop/ if not already present and give
your hostname as the entry.
```

FIGURE 13.8
Editing core-site.xml.

```
(F)Create a file by name "slaves" under $HADOOP_HOME/etc/hadoop/ if not already present and give
your hostname  as the entry

(G)Now format the namenode and start your cluster
$ hadoop  namenode  -format
$ start-all.sh
$jps (this should display all the 5 process )s

Open your browser and type the following address to view your NameNode and ResourceManager UI
respectively

<public_dns_name>:50070
<public_dns_name>:8088
```

FIGURE 13.9
Steps for configuration of Hadoop.

that depend on use of data, application, and requirement. It also tests Hadoop applications and unlike the stand-alone mode, HDFS is accessible in such an environment. Single-node clusters are good for development.

13.5.1 Importance and Benefits of Using Single-Node Hadoop Cluster Setup

The major benefits of this work are given below:

- Understanding is beneficial, so a single-node cluster is very important for new readers for learning and testing purposes.
- Due to the exponential growth of real-time data, parallel processing capabilities of a Hadoop cluster are used that will quicken the pace of analysis process. However, the handling and processing capabilities of a Hadoop cluster are depicted as unsatisfactory while growing the capacity of real-time data. To handle this situation in the application logic, the Hadoop clusters can be easily broadened to match with the speed of the analysis process by including added cluster nodes deprived of modifications.
- Cluster setup in Hadoop is an inexpensive process, as normally it is handled by cheap commodity hardware. However, a powerful Hadoop cluster can be set up without spending on high-priced server hardware.
- Hadoop clusters are robust to failure, i.e., data replication plays a crucial role. In case of failure of any node, the replicated copy can be used for the data analysis, which is stored on another node.

13.6 Multinode Cluster

In a multinode cluster, a master-slave architechture exists but there can be one or more slaves in it. For setting up the multinode cluster, the following steps are invloved:

Step 1 ensures the network condition by setting up the network.

Step 2 demonstrates the installation of Hadoop on the master. For this, certain entries are to be inserted to the host file, Java is required, therefore it is to be installed, and then configuration of ssh needs to be done. After the configuration of ssh, the configuration of Hadoop takes place.

Step 2 can be seen in Figure 13.10 through Figure 13.18.

Step 3 shows the installment of Hadoop on slaves. For this, entries are to be added in the host file along with the basic requirement of Java 8. Further, the process includes creating a tarball of configured files, copying the tarball on all slaves, and configured Hadoop is untar. This is shown in Figure 13.19 and Figure 13.20.

Step 4 is the main step, which comprises formatting the name node, starting the HDFS services, and starting the YARN services for the Hadoop cluster to run. Step 4 is demonstrated in Figure 13.21.

Step1: *Setting up the network*
> *Consider 192.168.0.1 IP address as master and 192.168.0.2 as slave*

Step 2: *Installing Hadoop on Master*
> *Here we are installing Hadoop on master node in the distributed mode.*

I. *Requirements for Hadoop 2.6 Multi Node Cluster Setup:*
Given below are the Hadoop installation requirements for a multi node cluster:

a. *Adding the Entries into the hosts file*
Add the following entries of master and slave:
1. sudo nano /etc/hosts
2. MASTER-IP master
3. SLAVE01-IP slave01
4. SLAVE02-IP slave02

FIGURE 13.10
Basic step towards setting up a multinode cluster.

---

b. *Installation of Java 8(Recommended Oracle Java)*
- Installing Python Properties

  sudo apt-get install python-software-properties
- Adding Repository

  sudo add-apt-repository ppa: webupd8team/java
- Updating the source list

  sudo apt-get update
- Installing Java

  sudo apt-get install oracle-java8-installer

FIGURE 13.11
Installation and configuration with Java.

---

c. *Configuration of SSH*
- Installing Open SSH Server-Client

  sudo apt-get install openssh-server openssh-client
- Generating Key Pairs

  ssh-keygen -t rsa -P ""
- Configuring passwordless SSH

  Copy the content of .ssh/id_rsa.pub (of master) to .ssh/authorized_keys (of all the slaves as well as master)
- Check by SSH to all the Slaves
1. ssh slave01
2. ssh slave02

FIGURE 13.12
Configuration and installation of SSH Hadoop.

> **III.** *Hadoop multi-node cluster setup Configuration*
> Setting up Hadoop configuration while installing Hadoop
> **a.** *Edit .bashrc*
> Edit .bashrc file located in user's home directory.
> Add the following environment variables:
> 1. export HADOOP_PREFIX="/home/ubuntu/hadoop-2.5.0-cdh5.3.2"
> 2. export PATH=$PATH:$HADOOP_PREFIX/bin
> 3. export PATH=$PATH:$HADOOP_PREFIX/sbin
> 4. export HADOOP_MAPRED_HOME=${HADOOP_PREFIX}
> 5. export HADOOP_COMMON_HOME=${HADOOP_PREFIX}
> 6. export HADOOP_HDFS_HOME=${HADOOP_PREFIX}
> 7. export YARN_HOME=${HADOOP_PREFIX}

FIGURE 13.13
Multinode cluster setup.

> **b.** *Check environment variables*
> Check the availability of environment variables in the .bashrc file:
> 1. bash
> 2. hdfs
>
> (the command not found error should not occur)
>
> **c.** *Editing the hadoop-env.sh*
> Edit configuration file hadoop-env.sh (located in HADOOP_HOME/etc/hadoop) and set JAVA_HOME:
> export JAVA_HOME=<path-to-the-root-of-your-Java-installation> (eg: /usr/lib/jvm/java-8-oracle/)

FIGURE 13.14
Checking environment variables and editing Hadoop-env.ssh file.

> **d.** *Editing core-site.xml*
> Edit configuration file core-site.xml (located in HADOOP_HOME/etc/hadoop) and add following entries:
> ```
> <configuration>
> <property>
> <name>fs.defaultFS</name>
> <value>hdfs: //master:9000</value>
> </property>
> <property>
> <name>hadoop.tmp.dir</name>
> <value>/home/ubuntu/hdata</value>
> </property>
> </configuration>
> ```

FIGURE 13.15
Editing core-site.xml.

```
e. Editing hdfs-site.xml
Edit configuration file hdfs-site.xml (located in HADOOP_HOME/etc/hadoop) and add following entries:
<configuration>
<property>
<name>dfs.replication</name>
<value>2</value>
</property>
</configuration>

f. Editing the mapred-site.xml
Edit configuration file mapred-site.xml (located in HADOOP_HOME/etc/hadoop) and add following
entries:
<configuration>
<property>
<name>mapreduce.framework.name</name>
<value>yarn</value>
</property>
</configuration>
```

FIGURE 13.16
Editing hdfs-site.xml and mapred-site.xml file.

```
g. Edit yarn-site.xml
Edit configuration file mapred-site.xml (located in HADOOP_HOME/etc/hadoop) and add following
entries:
<configuration>
<property>
<name>yarn.nodemanager.aux-services</name>
<value>mapreduce_shuffle</value>
</property>
<property>
<name>yarn.nodemanager.aux-services.mapreduce.shuffle.class</name>
<value>org.apache.hadoop.mapred.ShuffleHandler</value>
</property>
<property>
<name>yarn.resourcemanager.resource-tracker.address</name>
<Value>master: 8025</value>
</property>
<property>
<name>yarn.resourcemanager.scheduler.address</name>
<value>master:8030</value>
</property>
<property>
<name>yarn.resourcemanager.address</name>
<Value>master: 8040</value>
</property>
</configuration>
```

FIGURE 13.17
Editing yarn-site.xml.

h. *Editing the slaves*
Edit configuration file slaves (located in HADOOP_HOME/etc/hadoop) and add following entries:
1. slave01

2. slave02

FIGURE 13.18
Editing the slaves.

Step 3: *Installation of Hadoop On Slaves*
I. *Requirements for setup on all the slaves*
  • Add Entries in hosts file

  • Install Java 8 (Recommended Oracle Java)

II. *Copy the configured setups from master to all the slaves*
a. *Create tarball of configured setup*
tar czf hadoop.tar.gz hadoop-2.5.0-cdh5.3.2
(Run this command on Master)

FIGURE 13.19
Installing Hadoop on slaves.

b. *Copy the configured tarball on all the slaves*
scp hadoop.tar.gz slave01:~
(Run this command on Master)
scp hadoop.tar.gz slave02:~
(Run this command on Master)

c. *Un-tar configured Hadoop setup on all the slaves*
tar xzf hadoop.tar.gz
(NOTE: Run this command on all the slaves)
"Hadoop is set up on all the Slaves. Now Start the Cluster"

FIGURE 13.20
Copying tarball on all slaves and untar configured Hadoop on all the slaves.

Step 4: *For Starting the Hadoop Cluster*
Let us now learn how to start Hadoop cluster?

I. *Formatting the name node*
bin/hdfs namenode -**for**mat
(Run this command on Master)
(This activity should be done once when you install Hadoop, else it will delete all the data from HDFS)

II. *Starting the HDFS Services*
sbin/start-dfs.sh
(Run this command on Master)

III. *Starting the YARN Services*
sbin/start-yarn.sh

FIGURE 13.21
Initialization of Hadoop cluster using HDFS and YARN services.

13.7 Conclusion

This chapter covers the details of Apache Hadoop, which is able to import data into HDFS, export data from HDFS, compress data, and transform huge amounts of data. It provides scalable, reliable, and distributed environments. This chapter shows applicability of numerous tools of Hadoop, such as MapReduce, Hive, Pig, and Mahout. These tools are helpful for maintaining and monitoring the environment, handling MapReduce jobs, and read and write data in HDFS, and processing the data for visualization. The Hadoop environment also handles structured as well as unstructured data.

Various technologies from Amazon, IBM, Microsoft, and many more handle big data. The manuscript outlined sources of big data and issues. This chapter has given procedural key points for each tool along with a brief impact of single-cluster environments. It also demonstrated the difference between single and multicluster environments. Due to technological enhancements and real-time requirements, the big data Hadoop has high use in the future for data science due to data exploration with full datasets, mining larger datasets, and large-scale preprocessing of raw data as well as data agility. In the future, there is a need to focus on multinode cluster setups for handling big data. Further reading on this topic can be done from works by other authors [28–42].

Acknowledgments

We are thankful to the Galgotias University, Greater Noida for providing us an environment and resources during this work. Authors are highly grateful to Mr. Vijay Raja for availing the online resources at Cloudera.

References

1. Yang, Q. (2015). Introduction to the IEEE transactions on big data. *IEEE Transactions on Big Data, 1*(1), 2–15.
2. Yang, C., Huang, Q., Li, Z., Liu, K., & Hu, F. (2017). Big data and cloud computing: Innovation opportunities and challenges. *International Journal of Digital Earth, 10*(1), 13–53.
3. Srivastava, S., et al. (2016). Big data—An Emerging and Innovative Technology: Survey. In 2nd International Conference on Computational Intelligence and Communication Technology (CICT), pp. 1–6.
4. Rehioui, H., et al. (2016). DENCLUE-IM: A new approach for big data clustering. *7th International Conference on Ambient Systems, Networks and Technologies, Procedia Computer Science, Elsevier, 83*, 560–567.
5. Subramaniyaswamy, V., Vijayakumar, V., Logesh, R., & Indragandhi, V. (2015). Unstructured data analysis on big data using MapReduce. *2nd International Symposium on Big Data and Cloud Computing, Procedia Computer Science, 50*, 456–465.
6. Afrati, F. N., & Ullman, J. D. (2011). Optimizing multiway joins in a MapReduce environment. *IEEE Transactions on Knowledge and Data Engineering, 23*(9), 1282–1298.

7. Dongyao, W., Sherif, S., Liming, Z., & Huijun, W. (2017). Towards big data analytics across multiple clusters. In 17th IEEE/ACM International Symposium on Cluster, Cloud and Grid Computing (CCGRID), IEEE, pp. 1–6.
8. Reddy, Y. C. A., Viswanath, P., & Reddy, B. (2017). Semi-supervised single-link clustering method. In International Conference on Computational Intelligence and Computing Research (ICCIC), IEEE, pp. 1–6.
9. Rehm, F., Klawonn, F., & Kruse, R. (2006). Visualization of single clusters. *International Conference on Artificial Intelligence and Soft Computing, LNCS, 4029,* 663–671.
10. Yadav, D. K., & Singh, K. (2016). A combined approach of Kullback-Leibler divergence method and background subtraction for moving object detection in thermal video. *Infrared Physics and Technology, 76,* 21–31.
11. Sharma, L., Yadav, D. K., & Singh, A. (2016). Fisher's linear discriminant ratio based threshold for moving human detection in thermal video. *Infrared Physics and Technology, 78,* 118–128.
12. Lakhani, A., Gupta, A., & Chandrashekharan, K. (2015). IntelliSearch: A search engine based on big data analytics integrated with crowdsourcing and category-based search. In International Conference on Circuits, Power and Computing Technologies, pp. 1–6.
13. Sehrish, S., Mackey, G., Shang, P., Wang, J., & Bent, J. (2013). Supporting HPC analytics applications with access patterns using data restructuring and data-centric scheduling techniques in MapReduce. *IEEE Transactions on Parallel and Distributed Systems, 24*(1), 158–168.
14. Sharma, S. (2016). Expanded cloud plumes hiding big data ecosystem. *Future Generation Computer Systems, 59,* 63–92.
15. Sharma, S., Tim, U. S., Gadia, S., Shandilya, R., & Peddoju, S. (2014). Classification and comparison of NoSQL big data models. *International Journal of H. Haines, Introduction to HBase, the NoSQL Database for Hadoop.*
16. Kang, S. J., Lee, S. Y., & Lee, K. M. (2015). Performance comparison of OpenMP, MPI, and MapReduce in practical problems. *Advances in Multimedia,* 575687.
17. Jorge, L., Ortiz, R., & Anguita, D. (2015). Big data analytics in the cloud: Spark on Hadoop vs MPI/OpenMP on Beowulf. In INNS Conference on Big Data, pp. 121–130.
18. Simović, A. (2017). Recommender systems in the big data environment using Mahout framework. In 2017 25th Telecommunication Forum (TELFOR), IEEE.
19. Manogaran, G., & Lopez, D. (2018). Health data analytics using scalable logistic regression with stochastic gradient descent. *International Journal of Advanced Intelligence Paradigms, 10*(1/2), 118–132
20. Plase, D., Niedrite, L., & Taranovs, R. (2016). Accelerating data queries on Hadoop framework by using compact data formats. In 2016 IEEE 4th Workshop on Advances in Information, Electronic and Electrical Engineering (AIEEE), IEEE Conferences, pp. 844–849.
21. http://code.google.com/mapreduce-framework/wiki/MapReduce
22. http://www.dummies.com/programming/big-data/hadoop/hadoop-sqoop-for-big-data/
23. Li, Y., Fang, S., & Zhang, H. (2018). Panoramic synchronous measurement system for wide-area power system based on the cloud computing. In 2018 13th IEEE Conference on Industrial Electronics and Applications (ICIEA), IEEE Conferences, pp. 764–768.
24. https://www.ibm.com/analytics/hadoop/avro
25. https://www.developer.com/db/10-facts-about-hadoop.html
26. Sharma, P., Bhatnagar, V., & Mahajan, K. (2016). Analyzing click stream data using Hadoop. In 2016 Second International Conference on Computational Intelligence & Communication Technology (CICT), IEEE Conferences, pp. 102–105.
27. Khorshed, T., Sharma, N. A., Dutt, A. V., Ali, A. B. B., & Xiang, Y. (2015). Real time cyber attack analysis on Hadoop ecosystem using machine learning algorithms. In 2015 2nd Asia-Pacific World Congress on Computer Science and Engineering (APWC on CSE), IEEE Conferences, pp. 1–7.
28. Manwal, M., & Gupta, A. (2017). Big data and Hadoop—A technological survey. In 2017 International Conference on Emerging Trends in Computing and Communication Technologies (ICETCCT), IEEE Conferences, pp. 1–6.

29. http://www.dummies.com/programming/big-data/hadoop/hadoop-pig-and-pig-latin-for-big-data/
30. https://www.dezyre.com/article/difference-between-pig-and-hive-the-two-key-components-of-hadoop-ecosystem/79
31. https://hortonworks.com/apache/sqoop/
32. https://www.dezyre.com/hadoop-tutorial/flume-tutorial
33. https://hortonworks.com/apache/oozie/
34. https://www.edureka.co/blog/apache-oozie-tutorial/
35. http://cloudurable.com/blog/avro/index.html
36. https://hortonworks.com/apache/mahout/
37. https://www.ibm.com/developerworks/library/bd-yarn-intro/Figure3Architecture-of-YARN.png
38. https://ws1.sinaimg.cn/large/a6c1ce2agy1fkl94iyz47j20f60b1gmx.jpg
39. https://www.tutorialspoint.com/hadoop/hadoop_hdfs_overview.htm
40. Streamset. (2016). IoT Reference Architecture for Hadoop, pp. 1–8.
41. Vision Cloudera. (2016). Apache Hadoop—The data management platform for IoT. http://vision.cloudera.com/hadoop-the-data-management-platform-for-iot/
42. Real-time big data platform—Leverage real-time analytics and IoT integration to get insights faster than ever. https://www.talend.com/products/big-data/real-time-big-data/

Part 4

Challenging Issues

14

Block-Based Adaptive Learning Rate for Detection of Motion-Based Object in Visual Surveillance

Lavanya Sharma[1] and Pradeep K. Garg[2]

[1]*Amity Institute of Information Technology, Amity University, Noida, Uttar Pradesh, India*

[2]*Civil Engineering Department, Indian Institute of Technology, Roorkee, India*

CONTENTS

14.1 Introduction

The world faces several safety issues in current times. This has spurred rapid growth of requirements in visual surveillance systems or computer vision systems, which leads to large-scale research work in this active area. Over the last decennium, the area of safety has seen extensive growth for academics, researchers, and industry. The proliferation of static or fixed cameras in urban and rural regions is a clear indication of daily requirements for continuous monitoring or visualization. Automated surveillance is a very rigorous and efficacious area of research due to the rapid growth of several elicitable social events and sturdy terrorism activities [1, 2]. These kinds of systems are vital for security purposes in various settings, such as shopping complexes or malls, highways, defense borders, indoor and outdoor robotics, intelligent transportation system (ITS), and public or private spots.

The main job of well-grounded detection of a particular object from a dynamic or cluttered background scene in video data is the groundwork for several high-point real-time applications. The task is not yet fully resolved. Until now, object detection has faced many issues related to the tracking of objects, such as bootstrapping, camouflage, shadowing, noisy image, illumination changes (sudden or gradual), water rippling, and slow leafy movements [3–5]. The detection of an object of interest is a very simple step for image analysis in various real-time applications, such as human-machine interaction. However, in several cases it becomes very difficult to stratify the foreground (FG) or background (BG) pixels, which leads to misclassification of foreground pixels as background pixels. So, it becomes very important to classify pixels clearly as FG pixels or BG pixels because the rest of the steps highly depend on correct object detection. In the above scenario, a nonmoving camera with static background method can be used to classify pixels correctly. To overcome this problematic issue, some further modifications must be added to enhance the pixels' quality. This chapter mainly focuses on handling the following given challenging issues in indoor video sequences [3, 6, 7]:

- **Illumination variations:** Gradual or sudden lighting changes (such as camouflage, foreground aperture, light switch on or off, bootstrapping, and moved object).
- **Motion-based object:** Scenario where an object is moving (condition is not static). The segmentation of a motion-based object in a video frame sequence is based on its movement.

This chapter also presents an improved background subtraction method for motion-based object detection for visual security-based scenarios by enhancing the existing method for motion-based object detection in indoor video sequences with illumination variations, both in terms of qualitative and quantitative measurements. The proposed method handles both the challenges more precisely as compared to the state-of-the-art methods. The experimental analysis shows that this new approach performs well to overcome the problem of false pixel detection, and also results in better quality of detected objects. All the experiments are carried out on self-created indoor datasets, namely Dim Light Hall and Dim Light Dining Room. In both cases, lighting is not proper, thereby ensuring illumination invariance.

14.2 Literature Review

Numerous techniques have been proposed for the background subtraction technique [8], including dealing with complex or dynamic backgrounds including shadowing, illumination changes (sudden or gradual), slow leafy motions, and many more. Stauffer [2] presents an adaptive BG mixture approach for real traffic applications. He modeled each pixel as MOG and used EM approximation to update the model. To control the adaptive rate, a learning parameter is used, but this model is highly sensitive to noise and low convergence if Gaussian adopts any new cluster. To improve the convergence rate, Lee [3] proposed a technique based on the adaptive Gaussian mixture model (GMM). Later, this work was extended by several authors [3, 7, 9–11]. Haque et al. [5] improved the performance of Stauffer et al. [2] in terms of stability and quality of pixels but this technique fails to provide accurate results in highly dynamic scenes as it gives false-positive or false-negative alarms.

Hati et al. [12] presented a method based on intensity range to get BG pixel location to resolve illumination and motion-based issues in the BG scene. Zhou et al. [13] proposed a DECOLOR scheme to update BG regions. This method requires some initial parameters to be predefined in advance, such as number of Gaussians that will lead to false alarms because it becomes very difficult to model BG using some Gaussians (3 or 5). Haines et al. [14] proposed another efficient approach for BG distribution estimation (Dirichlet process-based GMM) to resolve the above issues of letters. Yadav et al. [15] also improved the efficacy of GMM [5] by implementing other distance-estimation approaches. Outcomes of post-processing techniques performed better than the previous author's work.

Yadav et al. [16] presented a GMM-based approach using a quasi Euclidean distance-based thresholding mechanism. Sharma et al. [17] presented an effective method using histogram-based adaptive learning for background modeling to detect motion-based objects in video surveillance. This method handles both illumination problems and moving environmental effects, which increases the true-positive rate.

14.3 Proposed Technique

The proposed technique is an improved method of Sharma et al. [18] to model multimodal domain components dynamically with nonmoving backgrounds. In this enhancement, firstly, the BG model of considered state-of-the-art methods was improved. Secondly, some post-processing tools are used to get improved quality of detected objects, in terms of quantitative performance metrics, including the number of frames executed per second. The proposed work shows an outstanding improvement in the performance analysis in dynamic environments, such as motion-based objects and illumination variations (gradual or sudden).

14.3.1 Development of a Background Model

A background model (BGM) is developed using some initial number of frames. This work uses its own dataset created using a normal phone with a 5-megapixel camera. So, initially the video sequence is converted into number of frames. For this purpose, one frame is read at a time and then the colored video is converted to gray-scale to obtain the number of rows and columns for further BG modeling using Equations 14.1 and 14.2. Here, 30 numbers of frames are considered for modeling a BG assuming that there is no moving object in the considered video sequence.

$$vid = mmreader\left(Dimlight\_Hall\right) \tag{14.1}$$

$$No.\ of\ Frames = vid.Number\ of\ Frames \tag{14.2}$$

$$Bg_n(x_i, y_i) = Bg_n - 1(x_i, y_i) + fg_n(x_i, y_i)/N \tag{14.3}$$

Here, Bg_n (x_i, y_i) represents intensity of a particular pixel p at a time t in the background and fg_n (x_i, y_i) represents intensity of a particular pixel p at a time t in the ith frame. N denotes the total number of frames used for BG modeling. The video sequence was done for Dim Light Hall with very low lighting conditions and Dim Light Dining Room with moderate lighting conditions.

14.3.2 Histogram for Model Updating

To classify the moving pixels, an adaptive threshold technique is used with a suitable learning rate. In order to update the learning at runtime, a histogram is computed for the classified pixels, as follows:

$$h(r_{kl}) = n_k \tag{14.4}$$

Where, L denotes number of possible intensity level, k represents the change of intensity level for given interval [0, G] and n_k represents the number of pixels with intensity level (r_k). Normalized histograms can be calculated dividing the $h(r_k)$ elements by the total number of pixels (n) in the frame [9], which is represented as follows:

$$p(rk) = \frac{h(rk)}{n} \tag{14.5}$$

$$= \frac{nk}{n} \tag{14.6}$$

for k = 1, 2, 3,…… L and $p(r_k)$ represents the probability density function intensity level r_k. The nature of variables is discrete in this case, so we assume $p_r(r_j)$, j = 1, 2, 3, 4,….L which represents the histogram (*His*) related with the intensity level of an input frame sequence. For summations of discrete quantities [9, 19, 20], the equalization transformation becomes:

$$s_k = T(_{rk}) \tag{14.7}$$

$$= \sum_{j=1}^{k} pr(rj) \tag{14.8}$$

$$= \sum_{j=1}^{k} nj / n \tag{14.9}$$

Here, s_k denotes the intensity value in the resultant frame sequence correction to value in the input frame sequence. Histogram equalization attains enhancement by level distribution of input frame sequences over a broad range of intensity levels. Let us suppose a continual intensity level that is normalized to the interval [0, 1], where r and z represent the intensity scale of input and output frame sequences. The $p_r(r)$ is the probability density function (PDF) of input level frames, and $p_z(z)$ is the specified probability density function of output frames [19, 21].

$$S = t(r) \int_{0}^{r} pr(w)dw \tag{14.10}$$

$$h(z) = \int_0^z pz(w)\,dw = s \qquad (14.11)$$

For a frame with intensity levels z with specified density, the given equation can be computed as follows:

$$Z = H^{-1}(S) = H^{-1}\left[T(r)\right] \qquad (14.12)$$

$T(r)$ can be calculated from the input frame by using Equation 14.12, transformed levels z can be calculated where PDF is the specified $p_z(z)$. In case of discrete variables, the inverse of H (histogram) exists only if $p_z(z)$ is well-grounded.

14.3.3 Threshold Generation

In this stage, a FG modeling method is investigated for the detection of motion-based objects in video sequences. In the testing stage, this method classifies the moving pixels with a suitable threshold, and also updates the BG using an appropriate learning rate ($\alpha = 0.01$). The learning rate is updated using a histogram of classified resultant frame sequences, and the BG model.

$$\text{Th} = \text{double} \left\{\left[f1(x_i, y_i) + \text{mean Value}\left(\text{Bg}_n(x_i, y_i)\right)\right] \times 6.2\right\}; \qquad (14.13)$$

$$\text{Hist\_sum}_n(x_i, y_i) = \left\{Fg_n(x_i, y_i) + Bg_n(x_i, y_i)\right\}; \qquad (14.14)$$

Where, $f1(x_i, y_i)$ is current frame, Bg_n is background modeled frame, and $\text{Hist\_sum}_n(x_i, y_i)$ denotes summation of resultant and BG model frame.

14.3.4 Motion-Based Pixel Classification

In order to classify the motion-based pixels correctly as FG or BG, the given classification technique is applied.

```
if (Diffₙ (x, y) ≥ Th)
{
                Fgₙ (x, y) = true; // pixel is in motion, i.e.,
                   part of foreground
    else
                Bgₙ (x, y) = false; // pixel is a part of
                   Background
} // end if
```

$$(14.15)$$

14.3.5 Maintenance of Learning Rate and Background Model at Runtime

To handle gradual illumination variations and motion in BG scenes, the BG model is updated at runtime for each forthcoming test frame sequence. In this method, learning rate α is used to overcome the complex or dynamic BG issues present in the BG scene through updating the BG model using Equation 14.3. The adaptive learning rate can be calculated using Sharma et al. [18, 21–23].

14.3.6 Foreground Enhancement

To overcome the above-mentioned issues, various morphological filters were used to handle the misclassifications of pixels. A structuring element (SE) with radius 2 has been applied, and holes have been filled by a hole-filling method of image processing. Due to rapid increases in the BG scene, some noisy pixels or blobs have been detected that were removed using the four connectivity-based component, where P2, P4, P6, P8 are immediate neighbors of pixel P, blob labeling, and thresholding technique. Finally, region-based thresholding has been applied to consider the isolated blobs as part of the BG scene that increases the true-positive rate and at the same time reduces the false-positive rate [19].

$$Fgframe = bwareaopen(Fgframe, P, conn); // \text{ removes all connected components} \quad (14.16)$$

Let us suppose P is a four-neighbor component, p_i and p_j are pixel pairs in P. Then, a sequence of pixels p_i, ..., p_j can exist if all the pixels present in the frame sequence in this given set are supposed to be black and secondly, every two pixels that are adjacent are four-neighbors in the given sequence.

14.4 Experimental Evaluation

Several experimental evaluations were carried out with this proposed approach on grayscale frame sequences. The comparative results were evaluated both qualitatively and quantitatively. Qualitatively was by means of visual examination, and quantitatively was by comparing the outcomes of the proposed work with other existing approaches using some quantitative parameters. In this evaluation, self-created indoor video datasets were used for BG subtraction. The first sequence, called *Dim Light Hall*, consists of 256 video frames, and the second sequence, *Dim Light Dining Room*, consists of 290 video frames with 430×240 frame size. In both scenarios, lighting is not proper, thereby ensuring illumination invariance. For experimental analysis, some predefined set of values with some perquisite requirements are taken into consideration as mentioned in Table 14.1. For this evaluation, ground truth images have been generated manually, and both the videos were recorded by a normal phone with a 5-megapixel camera. For qualitative analysis, some of the frames of both datasets are demonstrated with their ground truth mask, and the results are shown in Figure 14.1.

14.4.1 Qualitative Analysis

As shown in Figure 14.1, two video datasets have been used in this work for quality analysis. The visual analysis shows that the results of the proposed work outperforms the considered state-of-the-art method. In this work, five consecutive frame sequences have

TABLE 14.1

Prerequisites for Experimental Evaluations

Requirement	Description
Datasets	Indoor environment
Operating system	Windows 10
Processor	Intel (R) Core(TM) i5
Processing speed	CPU 1.8 Ghz
Storage	8 GB RAM
Tool used	MATLAB® 2012

been considered for benchmark evaluation. That frame sequence has illumination or lighting effects; due to this issue, handling illumination variation in indoor sequences is a tough task. Figure 14.1 shows the visual analysis for considered frame sequences [row-wise: (i) indoor datasets and their frame sequences, (ii) original frame, (iii) ground truth, (iv) proposed work, (v) Sharma et al. [18], and (vi) Haque et al. [5].

The proposed work shows better outcomes both qualitative and quantitative. As conventional background subtraction approaches are not suitable for challenging issues, such as gradual or sudden illumination variation, some part of the BG is falsely detected as FG to avoid these false classifications. Therefore, some improvements need to be incorporated.

14.4.2 Quantitative Analysis across Combination of Metrics

For quantitative analysis, few frames are displayed from two datasets. *Dim light Hall* has low lighting conditions, and *Dim Light Dining Room* has moderate lighting conditions along with the generated ground truth images. The outcome is presented in Figure 14.1. The system performance of some BGS approaches is evaluated by using several statistical metrics: accuracy, precision, recall, F1, fp_err, fn_err, and Total error [8, 23].

$$Precision = tp / (tp + fp) \qquad (14.17)$$

$$Recall = tp / (tp + fp) \qquad (14.18)$$

$$F - measure = 2*(precision*recall)/(precision+recall) \qquad (14.19)$$

$$Fp\_err = fp*100/rxc \qquad (14.20)$$

$$Fn\_err = fn*100/rxc \qquad (14.21)$$

$$Accuracy = (tp+tn)/(tp+tn+fp+fn) \qquad (14.22)$$

Where tp, fp, fn are total number of true-positive pixels, false-positive pixels, and false-negative pixels in motion mask. The Fp_err, Fn_err, r, c represent true-positive error, false-positive error, number of rows, and number of columns in a video frame, respectively. F-measure represents the harmonic mean of both precision and recall, which can be used for performance measure.

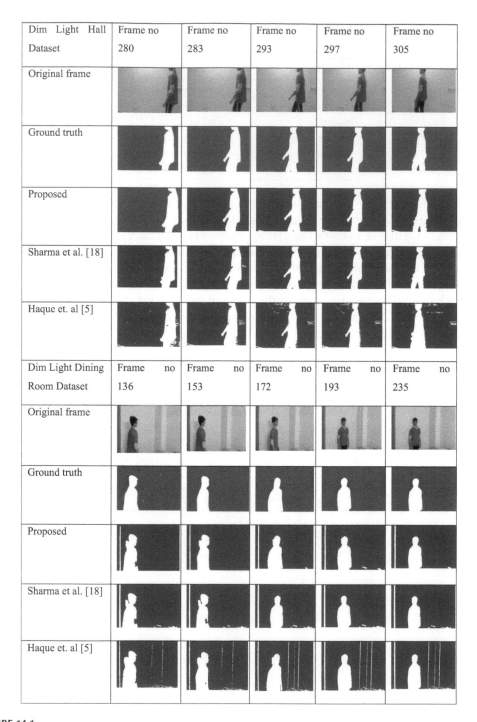

FIGURE 14.1
Qualitative results: row-wise (i) indoor datasets and their frame sequences, (ii) original frame, (iii) ground truth, (iv) proposed work, (v) Sharma et al. [18], and (vi) Haque et al. [5].

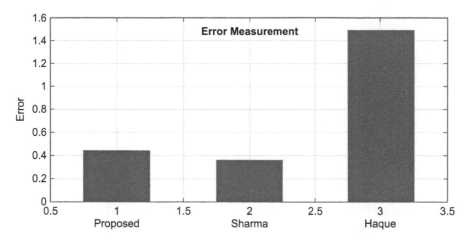

FIGURE 14.2
Percentage of average error analysis of all the competing methods.

Matthew's correlation coefficient (MCC) metric [24] is used for performance analysis of binary classifier, and can be expressed as:

$$MCC_{avg} = \frac{1}{n} \sum_n \int \frac{(tp.tn) - (fp.fn)}{\sqrt{(tp+fp).(tp+fn).(tn+fp).(tn+fn)}} \qquad (14.23)$$

Another performance measurement metric is percentage of correct classification (PCC_{avg}) [15, 20]. The PCC_{avg} for all video frame sequences can be represented as:

$$\int PCC_{av} = 100 \; x \; \frac{1}{n} \sum_n \frac{(tp+tn)}{(tp+fn+fp+tn)} \qquad (14.24)$$

The average error of the proposed and peer methods is shown in Figure 14.2. The error value of the proposed method on individual datasets is better but the average value of the proposed method is greater than Sharma et al. [18] and less than Haque et al. [5].

14.5 Discussions and Observations

Here we present a brief overview of the results and analysis of the proposed work and considered peer methods. The performance is shown in Table 14.2, Table 14.3, Table 14.4, Figure 14.2, Figure 14.3, and Figure 14.4. Regarding detected results, the performance is summarized as follows:

- Haque et al. [5] have generated good results; apart from these results this method also generates few false-positive and negative alarms. Due to the misclassification of pixels, i.e., false negatives, its performance degrades. According to the experimental results and analysis, its performance is poor on both the *Dim Light Hall* and the *Dim Light Dining Room* datasets, as depicted in Table 14.2 and Table 14.3. This method generates maximum error values as compared to considered peer methods.

TABLE 14.2

Quantitative Analysis: Precision-Recall, Error Measurement for Each Considered Methods

Method	Dataset	Precision	Recall	TP_ Error	FP_ Error	Total_ Error	Average Error
Proposed	Dim Light Hall	**1.0000**	**1.0000**	0.0000	0.0000	0.0000	0.4412
	Dim Light Dining Room	0.9147	**1.0000**	0.8825	0.0000	0.8825	
Sharma et al. [18]	Dim Light Hall	0.9999	0.9989	0.0019	0.0145	0.0164	0.3593
	Dim Light Dining Room	0.9308	1.0000	0.7022	0.0000	0.7022	
Haque et al. [5]	Dim Light Hall	0.9354	0.9969	0.8902	0.0395	0.9298	1.4901
	Dim Light Dining Room	0.8206	1.0000	2.0505	0.0000	2.0505	

TABLE 14.3

F-Measure, Accuracy, PCC, MCC, and Time Analysis for All Considered Methods

Method	Dataset	F-Measure	Accuracy	MCC	PCC
Proposed	Dim Light Hall	**1.0000**	1.0000	1.0000	0.0000
	Dim Light Dining Room	0.9554	0.9907	0.9514	0.9308
Sharma et al. [18]	Dim Light Hall	0.9994	0.9998	0.9993	0.0171
	Dim Light Dining Room	0.9642	0.9926	0.9608	0.7361
Haque et al. [5]	Dim Light Hall	0.9652	0.9901	0.9601	0.9948
	Dim Light Dining Room	0.9015	0.9775	0.8945	2.2451

TABLE 14.4

Time Analysis for Each Method over All Video Frame Sequences

	Datasets (frame size of each dataset: 432 × 240)	
Method	Dim Light Hall (256 frames)	Dim Light Dining Room (290 frames)
Proposed	27 frames per sec; 0.0375 sec/frame	19.737 = 20 frames per second 0.0494 sec/frame
Sharma et al. [18]	0.0402 sec/frame or 24 frames per second	0.0454 sec/frame or 22 frames per second
Haque et al. [5]	15.7375 frames per second; 0.0635 sec per frame	0.0688 second per frame; 14.5359 = 15 frames per second

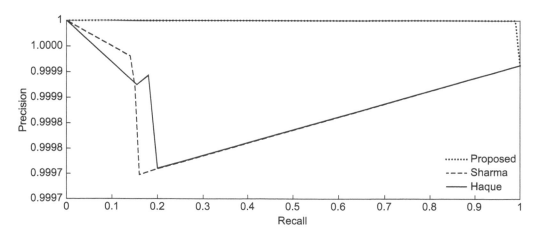

FIGURE 14.3
Precision-recall curve for all considered methods.

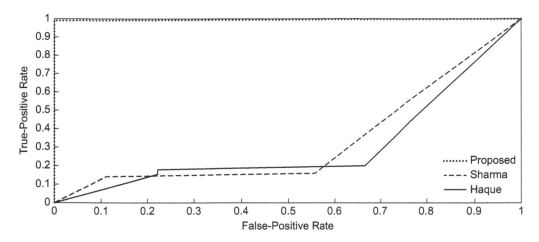

FIGURE 14.4
ROC-curve for proposed method with all considered state-of-the-art methods.

- Sharma et al. [18] have generated some false alarms. This method has good f-measure value. It also shows better accuracy as mentioned in Table 14.2. It shows good accuracy values on both the *Dim Light Hall* and the *Dim Light Dining Room* datasets.

- The proposed method depicts better performance. Its performance is excellent in the case of the *Dim Light Hall* dataset in terms of accuracy, MCC $_{(avg)}$, and PCC $_{(avg)}$ percentage of bad classifications. It is also better in terms of error and f-measure. The performance of the proposed work also compares with peer methods in terms of average value of the total error, MCC$_{(avg)}$ or PBC$_{(avg)}$. But this variation is much less, so the proposed method can be accepted with confidence.

The precision-recall curve is shown in Figure 14.3. According to this figure, the proposed method gives excellent performance on the *Dim Light Hall* dataset. At the top-right corner, the precision-recall curve is falling down and the whole curve depicts excellent results (dotted line curve) as compared to the Sharma et al. [18] (dashed line curve) and Haque et al. [5] (solid line curve). According to Figure 14.3, the proposed method gives better results as compared to the considered peer methods.

The curve in Figure 14.4 shows that the proposed method gives excellent performance on the *Dim Light Hall* dataset. At the top-left corner, the ROC-curve is below the boundary line of the top-most line and the whole curve depicts excellent results (dotted line curve) as compared to Sharma et al. [18] (dashed line curve) and Haque et al. [5] (solid line curve). So, according to Figure 14.4, the proposed work gives better results as compared to the considered literature for analysis.

14.6 Merits

The proposed work has some merits, which are given below:

- A remarkable advantage of the proposed work is that few external users' defined parameters are required as compared to the considered methods.

- Complexity is also very less than other methods, and well-suited for realistic scenarios because self-created indoor datasets were used in this work where lighting is not proper, thereby ensuring the illumination invariance.
- The proposed work is very effective and fast for background modeling, and computation time is also better than the other methods.
- The benchmark evaluation also demonstrates that the proposed work effectively reduces the illumination problem in indoor video sequences with low and moderate lighting conditions.
- Processing of frames per second is fast because this method works on the concept of frame by frame processing and not pixel by pixel. This work processes one frame at a time, which results in fast execution.
- Implementation of this work is simple, and can be easily implemented in a MATLAB® environment.
- This method is simple and applicable for visual surveillance applications.
- Both the learning rate and background model are updating on a runtime basis to avoid false classification of pixels.

14.7 Demerits

The proposed work has some demerits also, which are listed below:

- Due to the limitation of hardware and other resources, this work processed 27 numbers of frames per second for dataset *Dim Light Hall*, and 20 frames per second for dataset *Dim Light Dining Room*. Its execution speed is better than Sharma et al. [18] for *Dim Light Hall* and for both in the case of Haque et al. [5].
- The learning rate (α) is updated using summation of histogram of classified resultant frame and the BG model.
- This method is applicable only for motion-based objects at near distance from a still or fixed camera [25–34].

14.8 Conclusion and Future Scope

In this chapter, an efficient and robust method is proposed for motion-based detection of objects in cases of indoor video sequence with illumination variations with a nonmoving camera. The main advantage of the work is that it performs well for both scenarios with variant illumination backgrounds. The proposed method handles both the challenges more precisely as compared to the methods taken into consideration for benchmark evaluation. The experimental results demonstrate that this new method performs well to overcome the issue of false pixel detection, and also results in better quality of motion-based detected objects. All the experiments are carried out on self-created indoor datasets, namely, *Dim Light Hall* and *Dim Light Dining Room*. Due to its fast execution, this method is best suited for real-time indoor applications where videos are captured with a nonmoving or static camera under illumination variations. In the future, the work can be extended using Internet of Things (IoT) and cloud environments.

References

1. Irani, M., Rousso, B., & Peleg, S. (1994). Computing occluding and transparent motions. *International Journal of Computer Vision, 12*(1), 5–16.
2. Stauffer, C., & Grimson, W. E. L. (1999). Adaptive background mixture models for real-time tracking. *IEEE Computer Society Conference on Computer Vision and Pattern Recognition, 2,* 252.
3. Lee, D. S. (2005). Effective Gaussian mixture learning for video background subtraction. *IEEE Transactions On Pattern Analysis Machine Intelligence, 27*(5), 827–832.
4. Butler, D., Bove Jr., V. M., Sridha, S. (2005). Real-time adaptive foreground/background segmentation. *EURASIP, 2005*(14), 2292–2304.
5. Haque, M., Murshed, M., & Paul, M. (2008). On stable dynamic background generation technique using Gaussian mixture models for robust object detection. In 5th Int. Conference on Advanced Video and Signal Based Surveillance, IEEE, pp. 41–48.
6. Toyama, K., Krumm, J., Brumiit, B., & Meyers, B. (1999). Wallflower: Principles and practice of background maintenance. In International Conference on Computer Vision, pp. 255–261.
7. Piccardi, M. (2004). Background subtraction techniques: A review. *International Conference on Systems, Man and Cybernetics, IEEE, 4,* 3099–3104.
8. Jung, C. R. (2009). Efficient background subtraction and shadow removal for monochromatic video sequences. *IEEE Transactions on Multimedia, 11*(3).
9. Yu, L., & Qi, D. (2011). Hölder exponent and multifractal spectrum analysis in the pathological changes recognition of medical CT image. In 2011 Chinese Control and Decision Conference.
10. Barnich, O., & Droogenbroeck, M. V. (2011). Vibe: A universal background subtraction algorithm for video sequences. *IEEE Transactions on Image Processing, 20*(6), 1709–1724.
11. Ng, K., & Delp, E. J. (2011). Background Subtraction using a pixel-wise adaptive learning rate for object tracking initialization. *Visual Information Processing and Communication II, Proceedings of SPIE, 7882,* 1–9.
12. Hati, K. K., Sa, P. K., & Majhi, B. (2012). LOBS: Local Background Subtracter for video surveillance. In 2012 Asia Pacific Conference on Postgraduate Research in Microelectronics & Electronics (PRIMEASIA), pp. 29–34.
13. Zhou, X., Yang, C., & Yu, W. (2013). Moving object detection by detecting contiguous outliers in the low-rank representation. *IEEE Transactions on Pattern Analysis Machine Intelligence, 35*(3), 597–610.
14. Haines, T. S. F., & Xiang, T. (2014). Background subtraction with Dirichlet process mixture Model. *IEEE Transactions on Pattern Analysis Machine Intelligence, 36*(4), 670–683.
15. Yadav, D. K., Sharma, L., & Bharti, S. K. (2014). Moving object detection in real-time visual surveillance using background subtraction technique. In 14th International conference on Hybrid Intelligent Systems (HIS2014), IEEE, pp. 79–84.
16. Yadav, D. K., & Singh, K. (2015). Moving object detection for visual surveillance using quasi-Euclidian distance. IC3T-2015, LNCS. *Advances in Intelligent Systems and Computing Series, 381,* 225–233.
17. Sharma, L., Yadav, D., & Singh, A. (2016). Fisher's linear discriminant ratio based threshold for moving human detection in thermal video. *Infrared Physics & Technology, 78,* 118–128.
18. Sharma, L., Yadav, D. (2017). Histogram-based adaptive learning for background modelling: moving object detection in video surveillance. *International Journal of Telemedicine and Clinical Practices, 2*(1), 74–92.

19. Digital Image Processing applications. (2018). http://web.ipac.caltech.edu/staff/fmasci/home/astro_refs/Digital_Image_Processing_3rdEd_truncated.pdf

20. http://www.imageprocessingplace.com/downloads_V3/root_downloads/tutorials/contour_tracing_Abeer_George_Ghuneim/connect.html

21. Sharma, L., & Lohan, N. (2019). Performance analysis of moving object detection using BGS techniques in visual surveillance. *International Journal of Spatio-Temporal Data Science, 1*(1), 22–53.

22. Yadav, D. K., & Singh, K. (2019). Adaptive background modelling technique for moving object detection in video under dynamic environment. *International Journal of Spatio-Temporal Data Science, 1*(1), 4–21.

23. Motion based object detection using background subtraction technique for smart video surveillance. http://hdl.handle.net/10603/204721

24. Yadav, D. K., & Singh, K. (2015). Motion-based object detection in real-time visual surveillance system using adaptive learning. *Journal of Information Assurance and Security, 10*(2), 89–99.

25. Haque, M., Murshed, M., & Paul, M. (2008). Improved Gaussian mixture for robust object detection by adaptive multi-background generation. In 19th International Conference on Pattern Recognition.

26. Cai, J. (2008). A robust video-based algorithm for detecting snow movement in traffic scenes. *Journal of Signal Processing Systems, 56*(2–3), 307–326.

27. Li, D., Xu, L., & Goodman, E. D. (2013). Illumination-robust foreground detection in a video surveillance system. *IEEE Transactions On Circuits And Systems For Video Technology, 23*(10).

28. Bouwmans, T., Porikli, F., & Hoferlin, B. (2014). *A handbook on background modeling and foreground detection for video surveillance: Traditional and recent approaches, implementations, benchmarking and evaluation.* CRC Press, Taylor and Francis Group.

29. Sharma, L., Yadav, D. K., & Bharti, S. K. (2015). An improved method for visual surveillance using background subtraction technique. In 2nd International Conference on Signal Processing and Integrated Networks, IEEE, pp. 421–426.

30. Kim, W., & Kim, C. (2012). Background subtraction for dynamic texture scenes using fuzzy color histogram. *IEEE Signal Processing Letters, 19*(3), 127–130.

31. Chiranjeevi, P., & Sengupta, S. (2012). Robust detection of moving objects in video sequences through rough set theory framework. *Image and Vision Computing.*

32. Bhaskar, H., & Dwivedi, K. (2015). Autonomous detection and tracking under illumination changes, occlusions and moving camera. *Journal of Signal Processing, Elsevier, 117,* 343–354.

33. Chandrasekar, K. S., & Geetha, P. (2018). Moving object detection techniques in traffic surveillance: A review. In Second International Conference on Electronics, Communication and Aerospace Technology (ICECA), IEEE, Coimbatore, India, pp. 116–121.

34. Kim, C., Lee, J., Han, T., & Kim, U. (2018). A hybrid framework combining background subtraction and deep neural networks for rapid person detection. *Journal of Big Data,* 5–22.

15

Smart E-Healthcare with Internet of Things: Current Trends, Challenges, Solutions, and Technologies

Lavanya Sharma[1], Pradeep K. Garg[2], and Sunil K. Khatri[1]

[1]*Amity Institute of Information Technology, Amity University, Noida, Uttar Pradesh, India*

[2]*Civil Engineering Department, Indian Institute of Technology, Roorkee, India*

CONTENTS

15.1 Introduction

In today's world, Internet of Things (IoT) is the next emerging advancement in the field of Internet with wearable sensors. The term *Internet of Things* was suggested by MIT in the 1990s. This area mainly focuses on connecting objects with one another via sensor equipment, such as global positioning system (GPS), radio frequency identification (RFID), infrared (IR) sensor, and other devices working with the Internet to form a larger network, which is also known as a sensing network. IoT allows persons to remotely access and control things, such as medical equipment, connecting cars, agricultural machines, medication dispensing, condition of a patient in post-anesthesia care unit, poultry and farms, self-driving cars, room lighting through personal control, GPS smart sole, and many more [1–4]. IoT has emerged as one of the advanced research domains with great challenges and scopes of advancement in industries, academic, and research. IoT also generates several new possibilities that allow the development of an immense amount of strategies, devices, equipment, and real-time-based applications.

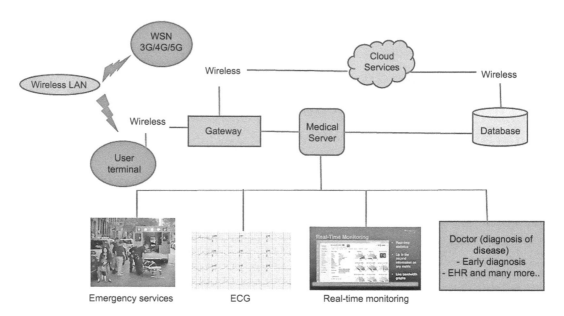

FIGURE 15.1
Recent trends in Internet of Things.

Nowadays, IoT is still more of an apparition than a realistic solution; it has yet to overcome many current challenging issues [5–8]. This area has a big scope for the technologies that will become a reality in forthcoming years. Various organizations, including industries and research labs, are already exploring new potentials that IoT brings to the next level of advancement in communication and technology. Figure 15.1 illustrates some key IoT trends, such as smart homes, grid, transportation, agriculture, industry automation, education, security, and healthcare [9–12]. Cost-effective interactions with secure data connectivity among healthcare centers, patients, and other clinical areas are an important aspect of today's realistic environment.

Modern healthcare networks using wireless devices to help in supporting the timely treatment of persistent diseases, early diagnosis of several chronic diseases, and real-time visualization of emergency cases. Over the last decade, this field has attracted attention from researchers and industry people who see its potential along with some of its open challenges. Currently, a number of prototypes, applications, and service providers are present in this field. In many countries across the world, organizations have developed policies to deploy this technology in medical emergencies and healthcare departments.

At this phase, a detailed understanding of ongoing research of IoT in conjunction with e-healthcare becomes fruitful to various researchers, academics, and stakeholders who are interested in further research in this field. This chapter provides a brief survey about the recent trends in healthcare research in context with IoT and uncovers various open issues that are present in this field, which must be addressed to transfigure the healthcare technology solutions provided by industry in this research area. In this regard, this chapter:

- Highlights the current trends and presents a brief summary of each of them.

- Provides a wide literature survey of IoT-based healthcare systems and their applications.

- Highlights the technical risk and current open challenges present in this field that must be addressed to make this technology more efficient and robust.
- Discusses the possible solutions provided by industry in the field in context with healthcare systems.
- Provides a case study based on a current solution, namely Pilot, and proposes a security model.
- Discusses core technologies, various policies, and strategies that support researchers in IoT innovation.

In fact, IoT has great impact when integrated with smart healthcare services, particularly e-health and emergency departments. In a survey conducted by KRC Research in the UK, Japan, and Germany this area represents more than 10% of the overall IoT market in terms of smart healthcare systems as shown in Figure 15.2. In the next decade IoT will generate $14.4 trillion in value across all industries. From the perspective of healthcare providers, this area helps them to improve patients' quality of life by adding various services that result in cost reduction and better user experiences. Considering the importance of emergency departments and applications in context of IoT, this chapter aims to presents a detailed overview of e-healthcare with IoT.

As per the McKinsey global report in the year 2013, the economic impact of IoT devices is going to increase from $3 trillion to $6 trillion by the end of the year 2025. Another report published by ABI Research also reveals that from the last year there was an enormous increase in the sales of leveraging mobiles, wearable medical devices, and sensor peripherals (used to gather critical data related to heart rate, body temperature, and barometer) as compared to the previous years and it is assumed that this will grow by more than 100 billion by the end of 2018, with revenues to exceed about $3 billion US dollars [13–16]. From the above two reports and a case study in India based on e-health also presented in this chapter, we conclude that there will be a very high impact on the worldwide economy, day-to-day life, and business [12] as shown

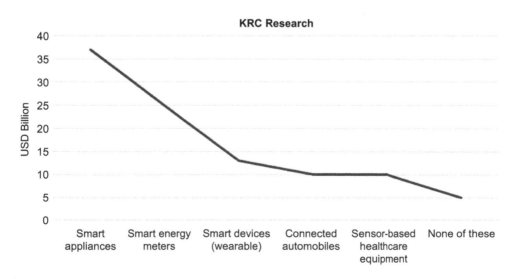

FIGURE 15.2
A survey conducted by KRC Research.

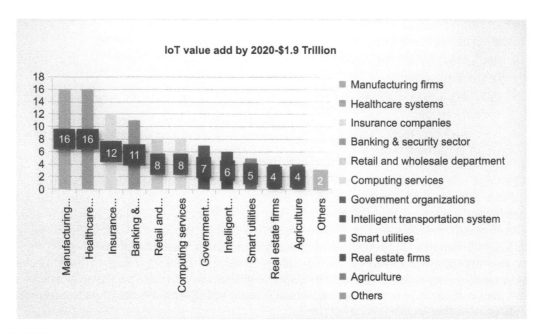

FIGURE 15.3
Economic impact of Internet of Things for 2020.

in the Figure 15.3. A large number of data or critical information will be transferred between wearable sensor-based devices and cloud or Internet services in the upcoming years.

This work mainly focuses on IoT's current challenging issues, applications, and solutions. This chapter also presents a framework, namely smart e-health care system. The rest of the chapter is ordered as follows. Section 15.2 reviews the literature of this technology for e-healthcare. Section 15.3 represents IoT health security. Application of e-healthcare in context of IoT is presented in Section 15.4. Section 15.5 deals with IoT and technologies of current trends. Section 15.6 outlines the available solutions to the e-health system. A case study in India based on e-health solutions provided to industries is presented in Section 15.7. In Section 15.8, a foresight on e-health trailblazers is described and discusses the unlimited opportunities in e-healthcare. In the last Section 15.9, the overall work of this chapter is concluded.

15.2 Literature Review

This section presents the related literature available during the last few years on IoT for healthcare. A cloud-based healthcare system that can be used for medical data storage is presented in which medical teams can access the gathered information using content service applications [2]. Babu et al. [6] proposed a remote health monitoring system based on the Open Geospatial Consortium (OGC) standard, a combination of wearable sensor- and web-based interfaces to measure several parameters such as body temperature and blood pressure of patients. In this work, sensors are used to collect the data and transmit the gathered information.

Serafim et al. [17] discuss an organization of networks that is used for monitoring patient health in rural regions with less population. Then, this collected data are transferred to the nearest healthcare center where they will be processed and easily accessible to the health-care department. This system helps emergency departments to analyze data and, if needed, also provide emergency medical assistance to the particular location. The correct diagnosis and patient monitoring rely only on the accurate analysis of the patient's medical records, which are recorded only occasionally. Dealing with data in terms of time management and quantity management makes analysis a little bit frustrating and results in clinicians that are prone to error. Even though the use of data mining and several visualization tools have been previously addressed as possible solutions to the above challenges [7, 18], these methods have gained a lot of attention in remote health monitoring systems in recent years [7, 13].

Various IoT-based devices are capable of sensing and exchanging critical data and then automatically transferring the collected data or information to the health department. These systems provide several important services such as an intrusion system to raise alarms to the nearby healthcare center in case of an emergency [4]. Remote healthcare monitoring can be also defined as a solution that collects the patient's physiologic data and then transmits these data from patients remotely. The basic architecture of this system mainly includes a user interface (i.e., smartphones, laptops, or tablets), data collectors (any kind of biosensors), and an Internet connection [14].

Islam et al. proposed a model to minimize the security risk and also discussed several inno-vations like big data and wearables, issues in healthcare and IoT policies, and their impact on society [19]. Khan [15] presents an effective healthcare monitoring system using IoT and RFID tags. This work shows robust results in a range of medical emergencies. In order to get the better results, the author used two phenomena to supervise and weigh the status of patients. The IoT healthcare platform is based on both a network platform and a computing platform. Jara et al. [20] present a service platform framework that focuses on residents' health informa-tion as shown in Figure 15.4. This framework represents a systematic hierarchical framework of four layers, where any concerned person or department can easily access the databases containing critical information from the application layer with the help of a support layer.

Another author proposed a similar architecture and designed a sensor-based network where nodes consist of ECG sensors, heart rate, and control actuators [21]. This system efficiently monitors human activities with the environment. The importance of standardizing interfaces across stakeholders of the IoT with healthcare towards the design of an open platform is high-lighted [20, 22, 23]. In this paper, three different categories of a cooperative ecosystem with standardization and without standardization are presented including e-health records (EHR), security measures, software and hardware interfaces as shown in Figure 15.5. Bazzani et al. [24] provide a solution using VIRTUS IoT middleware (an event-driven middleware package), which provides solutions to current IoT challenges, based on service-oriented architecture (SOA). Reliable and scalable communications of critical information over the Internet become more feasible even in cases of poor connectivity using this technique based on the XMPP.

15.3 IoT Healthcare Security

In recent years, IoT grows more rapidly day by day. It is also assumed that in upcoming years, healthcare departments are expected to adopt this technology and become more streamlined through new smart devices based on IoT and its applications. In today's

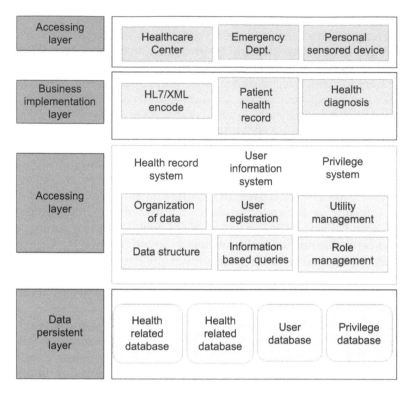

FIGURE 15.4
A framework: health information service model.

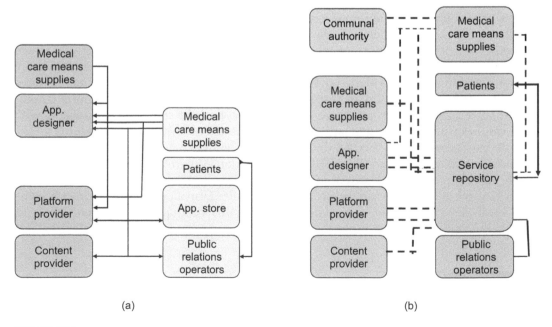

(a)   (b)

FIGURE 15.5
An ecosystem of organization (a) with regularization (b) without regularization.

world, healthcare devices and their related applications play a vital role in this field. With the increase of smart IoT-based devices for data sharing, the risk of attackers also increases. In order to facilitate this domain free from attackers or malicious users, one must be vigilant to analyze and identify distinct features of security and privacy including security requirements, threats model, and threat measures from a healthcare point of view.

a. **Security Requirements:** Security measures for IoT-based healthcare solutions are similar to the standard communications scenarios. In order to achieve safe and sound services, there is a need to focus on the security measures listed below.

- **Confidentiality:** This ensures the inaccessibility of critical information or data related to medical patients for malicious or unauthorized users. The messages resist revealing their confidential data to spoofers, eavesdroppers, and clickjackers.

- **Authentication:** It enables a smart device to ensure the identity of the agent with which it is communicating for secure communication.

- **Availability:** This ensures the availability of healthcare services to the authorized person or department in case of denial-of-service attacks.

- **Self-healing:** Smart sensor-based devices in this field sometimes can stop working or run out of energy. In this case, the other collaborating devices should facilitate a minimum level of security.

- **Resiliency:** If some interconnected smart device finds a medium in which to communicate, then a security mechanism should protect the devices, critical data, and network from any kind of attack.

- **Authorization:** This ensures that only authorized nodes can be medical patients, healthcare units, or associated agencies accessible for network-related services or other resources.

b. **Technical Risks and Current Challenges:** Several open challenging issues and technical risks are present in e-healthcare using IoT, which put it at risk of failure. Some of them are listed below:

- **Lack of electronic health record (EHR) system:** Sometimes the collected data smart device does not transfer to an electronic health record (EHR) system. In some cases, the EHR system is not centralized so it cannot be available to the clinician on time. While some EHR systems permit patients to import data into their record, this permission puts the providers into difficult situations as to how to handle the critical information of a patient who is outside of their records systems [11].

- **Interoperability:** Patients collect their data using different IoT devices for different purposes, like one for glucose-level collection and another for asthma-related symptoms. This means that the collected data may not reach the clinicians on time, the data may stay within the boundaries of each IoT system, and the data may not be visible to others. Due to lack of adequate interoperability, data collected from different devices may remain locked and result in loss of potential value to the team [11, 14].

- **Data security in e-healthcare:** At device level where data are collected and at the point of transmission of data until it reaches the final destination, securing that information is critical and required under the Health Information Portability and Accountability Act (HIPAA). Due to lack of security practices, various IT professionals have major concerns about the risks that are related to IoT device tampering [11].

- **Constant changes in hardware and connectivity technology:** To collect data, patients require more than one device, and in most cases the data are used along with a hub. At times, these hubs are not compatible with the devices and lack hardware components. Wireless connectivity devices such as Wi-Fi and Bluetooth can also force medical patients to have extensive hardware in their homes, which may be overwhelming and expensive [11].

- **Computational restrictions:** Processors of smart devices that are used for data collection are embedded with low speed and are not very powerful. These devices are not manufactured for computational operations; they can only act as actuators. A solution would be to minimize consumption of resources and maximize the outcomes in terms of security aspects [25].

- **Intelligent analysis and possible actions:** The last stage of implementation is extraction of valuable information from the collected data for analysis, but in some cases analysis is not up par due to flaws in the model or collected data, which leads to false alarms exposing limitations of the current algorithms. Sometimes the adoption process of new technology is very slow and situations are unpredictable [25–27].

- **Memory restrictions:** Most smart sensor-based devices used for data collection have low storage space. These smart devices are embedded with an operating system (OS) and system software. Therefore, built-in memories of such devices are not sufficient to execute the security protocols [28].

15.4 Application of E-Healthcare in Context of IoT

There are various applications of e-healthcare systems in context of IoT [10, 13, 14, 29–32]. Some of them are given below:

- **Real-time location (RTL) services:** Clinicals can use real-time location (RTL) services to easily track medical devices and equipment such as wheel chairs, nebulizers, and scales. Devices tagged with IP sensors can be located even when out of sight [13, 14].

- **Prediction of patient's arrival in post-anesthesia care unit (PACU):** Through IoT, medical staff and doctors can easily predict the time of arrival of patients to the PACU on an early basis. It also helps to monitor the present conditions of patients in real-time [10, 30].

- **Hand hygiene observance:** A survey about hospital infections was conducted by the Centers for Disease Control and Prevention (CDC), US; 1 out of every 20 patients gets and infection because of the unhygienic environment in hospitals.

Hand hygiene observance helps improve the hospital environment by storing the Identity, time, and location of healthcare staff, which can be accessed by the concerned department [10, 29, 32].

- **Reduce rates and improved patient stays:** Using IoT, healthcare staff access a particular patient's critical details from the cloud, which helps clinicians to provide medical care at affordable rates. IoT provides better patient stays in terms of immediate attention to patients in cases of emergencies, or also personal preferences such as room lighting via personal controls, and also communicating with family members [14, 32].

- **Remote monitoring system:** This system helps keep track of patients and healthcare staff. Sensors of this system can notify the concerned healthcare providers in case of emergency [32].

15.5 IoT-Based Technologies

Many technologies have the potential to modernize IoT-based healthcare services (Table 15.1). Some of them are listed below:

- **Big data:** This technology mainly includes large amounts of critical data that is generated from several medical sensor-based devices. It provides tools to improve accuracy of the diagnosis and to monitor patients efficiently.

- **Network:** Various wireless networks are used for transferring of critical data for short-range communications such as wireless personal area networks (WPANs), wireless body area network (WBANs), wireless sensor network (WSNs), wireless local area network (WLANs), 6LoWPANs (carry data packets in the form of IPv6 over IEEE 802.15.4 networks), and long-range communications such as cellular networks that are components of the physical infrastructure of any e-healthcare system. In addition, ultra-wideband (UWB), Bluetooth Low Energy (BLE), and RFID technologies are also used in designing the low-power-based devices and protocols for communications.

- **Cloud computing:** The combination of cloud computing and IoT in context with healthcare technologies provides services with worldwide access of sharing resources and also offers services based on Internet requests. Operations are executed to meet various requirements.

- **Wearables:** Several companies make wearable medical sensor-based devices, which make it quite easy for a person or medical patient to share information when connected over a network to the healthcare department. This results in timely diagnosis and treatment of problems.

- **Augmented reality (AR):** This technology play a very important role in IoT-based healthcare systems and also is very useful in the context of surgerical operations and remote monitoring. For example, AR can improve viewing a fetus inside a mother's womb, begin the process of treating cockroach phobia, and remind medical patients with AR glasses to take medications. Using this technology in combination with object detection in computer vision, a user becomes more interactive with the real world [22].

TABLE 15.1

IoT Firms with Their Future Vision

Organization	Future Vision
Samsung [33]	Samsung plans a strategic partnership to connect Samsung's ARTIK ecosystem with Philips Health Suite Digital Platform. This collaboration will ultimately allow the Samsung ARTIK ecosystem of connected devices to safely access and transfer critical data with Philips' cloud platform.
Philips [33]	The above collaboration of both the firms benefits healthcare. This collaboration can address the day-to-day requirements for connected health platforms to access, share, and analyze critical information that results in improvement of health systems. Healthcare providers can also achieve their goal of providing better care to medical patients in terms of diagnosis and treatment of disease.
IBM [34]	IBM in collaboration with other well-known firms across the world is set to develop smart sensor-based devices. It mainly focuses on emergency services such as connected home sensors, IoT data, and alerts that improve health, caregivers, and healthcare providers with tools to monitor care of the concerned person.
Intel [35]	Intel-powered smart devices can help in improving healthcare innovation. These include realistic analysis of data from edge to the cloud, enhancement of medical imaging and remote monitoring solutions, secure sharing of data between healthcare providers and patients, and many more. The Intel-based ecosystem with new technologies will result in improvements in terms of research and diagnosis.
Apple [36]	Apple Watch is an ultimate smart device, which collects biometric data and shares the information with providers. Basically it is considered a fitness tracker and a heart rate monitoring smart device that helps in improvement of secure data sharing, and healthcare gain.
Google [37]	This firm has opened its code for an open-source physical web service standard to arrange an easier approach to communicate with connected smart devices for medical purposes.
Microsoft [26]	Microsoft uses intelligent systems to provide a backbone of technologies in order to transform the medical care of patients. Maintenance of medical equipment, and the mode-concerned hospital or healthcare centers operate with innovation of IoT healthcare solutions. For example, BTT uses this technology (Microsoft-IoT) to monitor the critical activity of the human brain. Azure IoT wearables for cardio diagnosis help prevent premature death from heart-related disease.
Aeris [27]	This firm delivers IoT healthcare solutions for medical device manufacturers, remote monitoring of patients, and healthcare providers.

15.6 Available Solutions to E-Healthcare

Through IoT, researchers can easily analyze the accuracy of medical devices, and also shorten their day-to-day work. Using this technology, the healthcare industry will be able to provide the best possible solutions to their patients [10, 17, 19, 23, 31, 33, 38] as shown in Figure 15.6. However, this set of possible solutions does not yet include the IoT potential for an e-healthcare app:

- **Medication dispensing device by Philips:** For old-age patients so that they do not miss their doses [10].
- **Niox Mino by Aerocrine:** For asthma management, reads regular measurements of nitric oxide [17].
- **Bionexo:** This system provides various sets of solutions that allow real-time management of health supplies using Amazon Web Services [39].

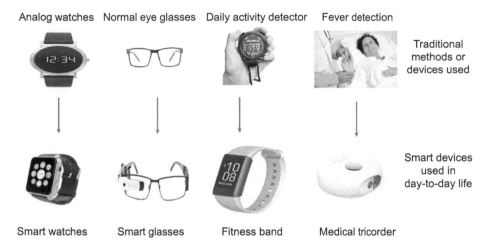

FIGURE 15.6
IoT healthcare products and prototypes.

- **UroSense by Future Path Medical:** To check a patient's body temperature and output of urine [11].
- **GPS SmartSole:** A wearable device used for dementia patients (patients having a habit of forgetting things) [10].
- **Gemed Oncologia:** It provides remote access to critical information via smart-phones and also helps in notifying the patient about appointments in advance [39].
- **DEV Tecnologia:** It comprises a cloud platform for device connectivity and management, which provides end-to-end IoT solutions offered by the company [40].
- **Period Tracker:** It predicts upcoming periods, tracks ovulation, and foretells fertility.
- **ElektorCardioscope:** It is potentially very useful to every person in terms of displaying the person's electrocardiograms (ECGs) on any smartphone or tablet.
- **Runtastic heart rate monitor:** This app measures the person's heartbeat using a smartphone or tablet camera fitted with sensors that help in getting a better accuracy rate and better sense of health and fitness.
- **Eye Care Plus:** This app checks the eyes' movements and tests to improve the quality of vision in a natural manner.
- **Asthma Tracker:** It track asthma-related symptoms in detail and helps in efficiently managing the dose of medication.
- **Fall Detector:** It monitors human activities very minutely and raises an alarm in case of a person falling.

15.7 Case Study in India

In literature various case studies are based on the current solutions of e-health (Figure 15.7). In India, a project named Pilot was started in Karnataka in December 2016. In a year in India about 2 million heart attacks are reported due to lack of medical aid. Death and

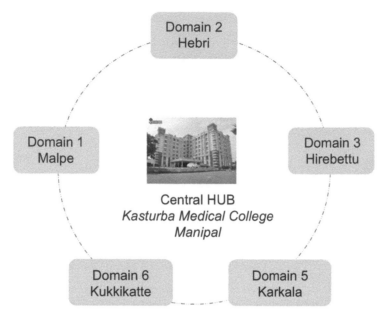

FIGURE 15.7
KMC Manipal hub.

disability due to heart attacks is a big health issue in India and a majority of the victims are youngsters [9].

The Philips Efficia ECG100 solution is a minute and transportable ECG machine-based mobile device, irrespective of any dependency on its infrastructure (Wi-Fi). Printing can be accessed remotely over the Internet, making the task simple and allowing earlier diagnosis and timely reporting of critical patient information to the hospitals. This system results in saving heart attack patients, as shown in Figure 15.8.

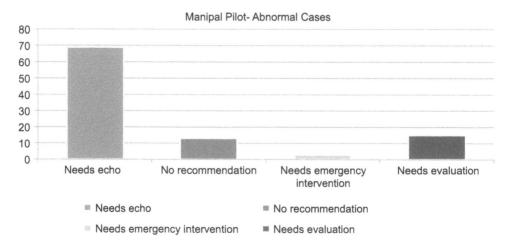

FIGURE 15.8
Manipal CPC pilot result.

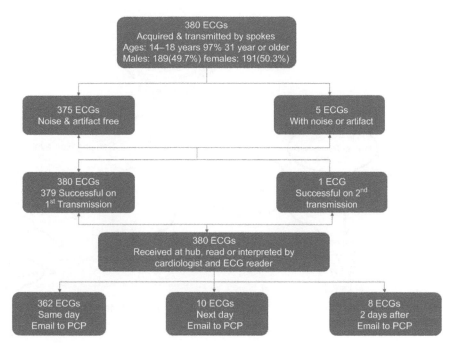

FIGURE 15.9
Critical data received at Manipal hub.

The above solution also empowers doctors at remote clinics, paramedics in ambulances, or non-cardiologists to make decisions on the basis of guidance provided by cardiologists at percutaneous coronary intervention (PCI) hospitals as shown in Figure 15.9. This also helps in improving care and outcomes of patients with severe chest pain throughout the area around a PCI hospital. It helps cost reduction, mental stress, and avoidable admissions of patients with low-risk chest pain symptoms [9].

15.8 E-Healthcare: Opportunities Unlimited

The section describes the working of the AI/block chain/IoT-based healthtech system. The e-healthcare system will allow patients, with wearables and portable medical devices (such as Implantable Medical Transceivers ZL70103), to automatically send their critical information using IoT from one place, which might be a home, to the concerned hospital authorities for medical diagnosis and treatment, to the clinical agencies. Even in cases of emergency, the information can be securely forwarded to the hospital authorities, including doctors and nursing staff, where the day-to-day data of patients, along with their personal information, can be stored at a permanent repository, that is, a medical server. With the e-healthcare system, the present condition of a patient can also be monitored as shown in Figure 15.10. This system offers great amenities to both medical patients and healthcare providers. This reduces delays for medical treatment after the diagnosis of problem. From the healthcare providers' point of view, after receiving a patient's critical data, they can start the appropriate treatment, which automatically results in saving of medicinal resources [11, 40–42].

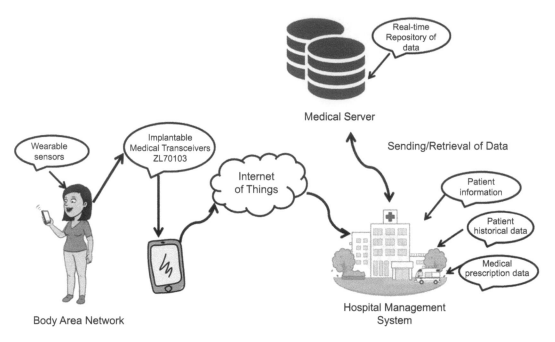

FIGURE 15.10
Framework of e-healthcare system in context with IoT.

Additionally, deep-tech interventions have various applications in healthcare, such as portability and remote monitoring of IoT-based smart sensors and integration with healthcare devices. This has the potential of not only keeping patients safe and healthy, but also of improving doctors' practice. IoT-integrated healthcare can also improve engagement and satisfaction of patients by allowing them to spend more time interacting with their doctors.

However, IoT-integrated healthcare does come with its fair share of challenges. The IT departments of hospitals will need to strengthen their IT infrastructure to accommodate the number of connected devices and the tremendous amount of data they bring in. The question of data security will also need to be addressed, especially if personal information is being shared with other connected IoT devices. Some other challenges of IoT in healthcare are managing multiple connected devices and a lack of integration with current industry-standard medical devices and interoperability of systems.

Nevertheless, in the next 5 years, there will be a massive increase in IoT-integrated healthcare, both on the doctors' side as well as at the back end. Some of the applications of IoT-integrated healthcare can be:

- Integrating technology with healthcare for medical devices gives people an opportunity to integrate other fitness devices, such as Fitbits, to bring patient-provided data into the cycle of care delivery.
- E-healthcare is also looking into glucometers, blood pressure monitors, and other devices that can collect data and statistics automatically and apply Emergency severity index ESI rules to allow doctors and the support team to mediate on the patient's vital signs.

- At-home, do it yourself (DIY) medical devices can be enabled to work on a wireless network to collect the data directly into the digital drives, increasing the workflow efficiency and ease in documentation.
- Live monitoring of health of infants in the neonatal intensive care unit (NICU).
- Artificial intelligence can be used to track ever-increasing inventory in areas, such as pharmacy, and overall inventory control in warehouses in a widespread manner. Medical institutions struggle to deploy strategies as adopted in the retail sector, using various tools and integrating them with currently used ERP solutions to manage the back end [41, 42].
- Taking forward the concept of RFID, adoption of IoT-integrated healthcare technology has not happened as rapidly as expected. However, the decreasing unit economics of the devices and the increasing reliability of the wireless infrastructure will help in driving this.
- By using the wireless infrastructure and tag devices such as wristbands and ID badges, hospitals can better manage foot traffic. IoT will also allow hospitals to analyze foot traffic, understand current holdups, and subsequently work to resolve these holdups. This will allow optimizing of the results and delivering value in a dynamic way rather than the old-fashioned static ways [43–86].

15.9 Conclusion

As healthcare is considered one of the most important sectors for tech-based applications, this chapter describes the current state. Also, the chapter predicts future steps for monitoring of health remotely using wearable sensors, implantable medical transceivers, and many other devices that improve healthcare even in cases of emergencies. This proposed framework of an e-health system helps in monitoring patient health on a regular basis and forwarding the critical information to the hospital doctors for immediate treatment. These solutions will also help researchers to gather monitoring data of patient health on a regular basis. Diagnostic reports showing the success or failure of current treatments could help improve the healthcare of other people with similar health problems. We will witness massive integration of technology with healthcare over the next few years. Hospitals have just started to explore the basics. Later, this technology can bring in a goldmine of data that can add tremendous value to many allied industries and create a snowball effect of value generation for all stakeholders.

References

1. Wei, L., Kumar, N., Lolla, V., & Keogh, E. (2005). A practical tool for visualizing and data mining medical time series. In Proceedings 18th IEEE Symposium on Computer-Based Medical System, pp. 341–346.
2. Rolim, C., & Koch, F. (2010). A cloud computing solution for patient's data collection in health care institutions. In Second International Conference on eHealth, Telemedicine, and Social Medicine, ETELEMED, pp. 95–99.

3. Mao, Y., Chen, Y., Hackmann, G., Chen, M., Lu, C., Kollef, M., & Bailey, T. C. (2011). Medical data mining for early deterioration warning in general hospital wards. In IEEE 11th International Conference on Data Mining Workshops (ICDMW), pp. 1042–1049.

4. Bui, N., & Zorzi, M. (2011, October 26-29). Health care applications: A solution based on the internet of things. In Proceeding of the 4th International Symposium on Applied Sciences in Biomedical and Communication Technologies, New York, NY, pp. 1–131.

5. Yang, N., Zhao, X., & Zhang, H. (2012). A noncontact health monitoring model based on the Internet of Things. In Proceeding 8th International Conference Natural Comput. (ICNC), pp. 506–510.

6. Babu, S., Chandini, M. K., & Vaidehi, V. (2013). Cloud-enabled remote health monitoring system. In International Conference On Recent Trends in Information Technology (ICRTIT), pp. 702–707.

7. Ukis, V., Balachandran, B., Rajamani, S., & Friese, T. (2013). Architecture of cloud-based advanced medical image visualization solution. In IEEE International Conference on Cloud Computing in Emerging Markets (CCEM), pp. 1–5.

8. Jin, J., Gubbi, J., Marusic, S., & Palaniswami, M. (2014). An information framework for creating a smart city through Internet of Things. *IEEE Internet of Things Journal, 1*, 112–121.

9. Case study of e-health. (2018). http://www.theiet.in/IoTPanel

10. Solutions to ehealth. (2018). https://www.cabotsolutions.com/2016/02/applications-iot-healthcare-industry

11. Challenging issues related to ehealthcare. (2018). https://www.sitepoInternationalcom/4-major-technical-challenges-facing-iot-developers/

12. Vasanth, K., & Sbert, J. (2014). Creating solutions for health through technology innovation. Texas Instruments. http://www.ti.com/lit/wp/sszy006/sszy006.pdf

13. Rao, B. (2013, June). The role of medical data analytics in reducing health fraud and improving clinical and financial outcomes. In Computer-Based Medical Systems (CBMS), 2013 IEEE 26th International Symposium, p. 3.

14. Kabir, H., Hossain, M., & Kwak, K. (2018). The Internet of Things for health care: A comprehensive survey. http://ieeexplore.iee.org/document7113786/

15. Khan, S. F. (2017). Health care monitoring system in Internet of Things (IoT) by using RFID. In 6th International Conference on Industrial Technology and Management (ICITM), pp. 7–10.

16. IoT based research. (2018). https://venkat-alagarsamy.blogspot.com/2014/12/internet-of-things-iot-next-revolution.html

17. Serafim, E. (2014). Uma Estrutura de Rede Baseada em Tecnologia IoT para Atendimento Médico em Áreas Urbanas e Rurais [dissertation]. Campo Limpo Paulista: Faculdade Campo Limpo Paulista.

18. Rios, T. S., & Bezerra, R. M. S. (2015, June 17-20). WHMS4: Um Modelo Integrado para Monitoramento Remoto de Saúde. In 10ª Conferencia Ibérica de Sistemas y Tecnologías de Información, Águeda, Portugal.

19. Riazul Islam, S. M., Kwak, D., Kabir, M. H., Hossain, M. S., & Kwak, K. S. (2015). The Internet of Things for health care: A comprehensive survey. *IEEE Access, 3*, 678–708.

20. Jara, A. J., Zamora-Izquierdo, M. A., & Skarmeta, A. F. (2013). Interconnection framework for mHealth and remote monitoring based on the Internet of Things. *IEEE Journal of Select Areas in Communications, 31*(9), 47–65.

21. Rasid, M. F. A., Kadir, N. A. A., Musa, W. M. N. M. W., & Noor, A. M. (2014). Embedded gateway services for Internet of Things applications in ubiquitous healthcare. In Proceeding 2nd International Conference Information and Communication Technology (ICoICT), pp. 145–148.

22. Vazquez-Briseno, M., Navarro-Cota, C., Nieto-Hipolito, J. I., Jimenez-Garcia, E., & Sanchez-Lopez, J. D. (2012). A proposal for using the Internet of Things concept to increase children's health awareness. In Proceeding 22nd International Conference Elect. Communication Computing (CONIELECOMP), pp. 168–172.

23. Augmented reality. (2018). https://en.wikipedia.org/wiki/Augmented_reality#Emergency_management/search_and_rescue

24. Bazzani, M., Conzon, D., Scalera, A., Spirito, M. A., & Trainito, C. I. (2012). Enabling the IoT paradigm in e-health solutions through the VIRTUS middleware. In Proceeding IEEE 11th International Conference Trust, Security Privacy Computing Communication (TrustCom), pp. 1954–1959.

25. Riazul Islam, S. M., Kwak, D., Humaun Kabir, M. D., Hossain, M., & Kwak, K-S. (2015). The Internet of Things for health care: A comprehensive survey. *IEEE Transaction, 3*, 678–708.

26. Microsoft healthcare solutions. (2018). https://www.microsoft.com/en-in/internet-of-things/customer-stories-healthcare

27. Delivering improved health outcomes and reduced cost of medical care with remote patient monitoring. (2014). http://www.aeris.com/for-enterprises/healthcare-remotepatient-Monitoring

28. Wearable devices. (2018). https://volersystems.com/v-2014/198-innovation-trends-for-sensors-and-wearable-devices/#iLightbox

29. Höller, J., Mulligan, C., Tsiatsis, V., & Karnouskos, S. (2014). *From machine-to-machine to the Internet of Things: Introduction to a new age of intelligence.* Amsterdam, The Netherlands: Elsevier.

30. Kaleem, U., Ali, M., & Zhang, J. (2016). Effective ways to use Internet of Things in the field of medical and smart health care. In 2016 International Conference on Intelligent systems Engineering.

31. Survey on IOT. (2018). https://www.analyticsvidhya.com/blog/2016/08/10-youtube-videos-explaining-the-real-world-applications-of-internet-of-things-iot/

32. Internet world stats. (2018). http://www.internetworldstats.com/top20.html

33. Samsung in collaboration with Philips. (2018). https://www.samsung.com/us/ssic/press/philips-and-samsungteam-to-expandconnected-health-ecosystem/

34. IBM healthcare solutions. (2018). https://www.ibm.com/internet-of-things/spotlight/iot-zones/iot-home

35. https://www.intel.in/content/www/in/en/healthcare-it/transforming-healthcare.html

36. Apple healthcare solutions. (2018). https://internetofthingsagenda.techtarget.com/feature/Healthcare-IoT-Apple-Watch-ready-to-change-patient-care

37. Physical web: Walk up and use anything. (2014). http://google.github.io/physical-web

38. Advances in intelligent systems and computing. (2018). http://link.springer.com/chapter/10.1007/978-3-319-31307-8_5

39. Zhao, W., Wang, C., & Nakahira, Y. (2011). Medical application on internet of things. In IET International Conference on Communication Technology and Application (ICCTA 2011), pp. 660–665.

40. Bui, N., & Zorzi, M. (2011). Health care applications: A solution based on the internet of things. In Proceeding of the 4th International Symposium on Applied Sciences in Biomed. and Communication Technology, pp. 1–131.

41. Sharma, L., & Lohan, N. (2019). Performance analysis of moving object detection using BGS techniques in visual surveillance. *International Journal of Spatio-Temporal Data Science, 1*(1), 22–53.

42. Sharma, L., & Yadav, D. (2017). Histogram-based adaptive learning for background modelling: Moving object detection in video surveillance. *International Journal of Telemedicine and Clinical Practices, 2*(1), 74–92.

43. Lan, M., Samy, L., Alshurafa, N., Suh, M. K., Ghasemzadeh, H., MacAbasco-O'Connell, A., & Sarrafzadeh, M. (2012). Wanda: An end-to-end remote health monitoring and analytics system for heart failure patients. In *Proceeding of the Conference on Wireless Health*, New York, NY, pp. 1–9.

44. Ray, P. (2014). Home health hub internet of things (H3IoT): An architectural framework for monitoring health of elderly people. In International Conference on Science Engineering and Management Research (ICSEMR), pp. 1–3.

45. Xu, B., Xu, L. D., Cai, H., Xie, C., Hu, J., & Bu, F. (2014). Ubiquitous data accessing method in IoT-based information system for emergency medical services. *IEEE Transaction Industrial Information, 10*(2), 1578–1586.

46. Yang, L., Ge, Y., Li, W., Rao, W., & Shen, W. (2014). A home mobile healthcare system for wheel-chair users. In Proceeding IEEE International Conference Computing Supported Cooperative Work Design (CSCWD), pp. 609–614.
47. Pang, Z., Tian, J., Chen, Q., & Zheng, L. (2013). Ecosystem analysis in the design of open platform-based in-home healthcare terminals towards the Internet-of-Things. In Proceeding International Conference Advance Communication Technology (ICACT), pp. 529–534.
48. Fan, Y. J., Yin, Y. H., Xu, L. D., Zeng, Y., & Wu, F. (2014). IoT-based smart rehabilitation system. *IEEE Transactions on Industrial Information, 10*(2), 1568–1577.
49. Jia, X., Chen, H., & Qi, F. (2012). Technical models and key technologies of e-health monitoring. In Proceeding IEEE International Conference e-Health Network, Appl. Services (Healthcom), pp. 23–26.
50. Miori, V., & Russo, D. (2012). Anticipating health hazards through an ontology-based, IoT domotic environment. In Proceeding 6th International Conference Innovation, Mobile Internet Services Ubiquitous Computing (IMIS), pp. 745–750.
51. Xu, B., Xu, L. D., Cai, H., Xie, C., Hu, J., & Bu, F. (2014). Ubiquitous data accessing method in IoT-based information system for emergency medical services. *IEEE Transaction Ind. Information, 10*(2), 1578–1586.
52. Zhang, X. M., & Zhang, N. (2011). An open, secure and flexible platform based on Internet of Things and cloud computing for ambient aiding living and telemedicine. In Proceeding International Conference Computing Manage (CAMAN), pp. 1–4.
53. Microfluidics based point-of-care platform for immunodiagnostics. (2018). https://achiralabs.com/acix100/
54. Blood bag monitoring device. (2018). http://bagmo.in/about.php
55. The Cellworks Biosimulation Platform. (2018). https://cellworks.life/technology
56. VAPCare: AI based secretions and oral hygiene management device. (2018). https://www.coeo.in/
57. A platform technology for healthcare solutions using 3D printed medical models. (2018). http://www.df3d.com/
58. A fetal monitoring device to prevent still births. (2018). https://www.empathydesignlabs.com/
59. Noxeno, which is a novel, dedicated, easy to use device for effective removal of impacted foreign bodies (NFB) from the nasal cavity. (2018). http://innaccel.com/
60. Soft tissue core biopsy (BioScoop™) and BxSeal™ technology. (2018). http://www.indiolabs.com/technologies.html
61. Developing solutions in the area of animal and human cell type identification. (2018). http://jivasci.com/
62. Developing immersive medical simulation technologies. (2018). http://www.mimyk.com/
63. Strip-based biosensors for the detection of E. coli in food and clinical samples. (2018). www.moduleinnovations.com/?page_id=177
64. Symptomatic uterine fibroids treatment. (2018). http://nesamedtech.com/#aboutus
65. Diagnostic platform for the miR-based detection of pre-eclampsia. (2018). http://prantaesolutions.com/
66. An imaging-based point-of-care diagnostic device. (2018). http://sminnovations.in/#portfolio
67. Atomic Force Microscope and micro-fluidic technologies for isolating single cells for applications in life sciences. (2018). http://shilpsciences.com/
68. Diagnostic pacifier which has been developed as a platform for saliva based screening of various neonatal health ailments. (2018). http://www.spotsense.in/
69. Clinical genomics and hepatotoxicity prediction platforms. (2018). https://strandls.com/strand-clinical-diagnostics/
70. Biomedical and bioimaging. (2018). http://unilumen.in/bio-photonics/
71. To monitor mental health and manage epilepsy. (2018). http://www.teblux.com/tjay.html
72. Diabetic peripheral neuropathy screening device. (2018). http://www.yostra.com/
73. A trypsin clearance assay kit. (2018). http://www.affigenix.com/internalresearchfocus.php

74. Nanotechnology based therapeutics for lysosomal storage disorders and other rare diseases. (2018). http://www.atenporus.com/intellectual-property/
75. Treatment of antibiotic resistant bacterial infections. (2018). http://bugworksresearch.com/elude.html
76. Age-related macular degeneration (AMD). (2018). http://eyestem.com/science/
77. A platform for DNA synthetic vaccines. (2018). http://www.geniron.in/
78. Development of recombinant proteins and monoclonal antibodies against clinically relevant targets through the integration of bioinformatics, recombinant DNA technology. (2018). http://gng.asia/gng/table.php
79. Nano-magnetic discs for ablation of cancer cells in vitro. (2018). http://nurtureearth.in/TechnologyDevelopment%20.aspx
80. Novel vaccine molecules against fatal human diseases. (2018). http://pentavalent.in/about_us.html
81. Technology platforms with the vision to manufacture personalized homo-chippiens and human organs on demand. (2018). http://pandorumtechnologies.in/science-tech/
82. Biologics and biobetters. (2018). http://theramyt.com/
83. Stable transgenic cell lines using CRISPR-Cas9 based genome editing and viral vectors. (2018). http://viravecs.com/
84. Treatment modalities for cancer patients. (2018). http://wrbio.com/?page_id=17
85. Ayurveda's herbal tradition into a range of proprietary technologies. (2018). http://www.robustherbals.com/
86. A novel air purification device for the reduction of airborne bacteria. (2018). http://biomoneta.com/

16

A Foresight on E-Healthcare Trailblazers

Lavanya Sharma[1], Pradeep K. Garg[2], Naman Kumar Agrawal[3], and Himanshu Kumar Agrawal[4]

[1]*Amity Institute of Information Technology, Amity University, Noida, Uttar Pradesh, India*

[2]*Civil Engineering Department, Indian Institute of Technology, Roorkee, India*

[3]*Young Professional, Atal Innovation Mission, NITI Ayog, New Delhi, India*

[4]*Startup Fellow, Ministry of Human Resource Development (MHRD) Innovation Cell, AICTE, New Delhi, India*

CONTENTS

16.1 Introduction

Health is one of the most important things in life. In one of the largest organization deals with technological developments, IBM invested about $4 billion in medical technologies, with the hopes that IoT will improve emergency department care for patients and also relieve pressure on staff in medical health centers. Over the last few years, the cost of sensory devices has decreased by half, 60 times less than processing costs and 40 times less than bandwidth costs. Using recent developments, organizations are connecting mobile devices and sensors to human bodies in order to accumulate and evaluate huge amounts of information (one person's e-health-related data are equivalent to approximately 300 million e-books). Organizations are also using IoT to find new ways to diagnose, prevent, and treat medical problems. The IoT brings many benefits to e-healthcare departments: reduction in errors; new insights from conventional methods to recent ones; betterment of medical centers in terms of workload, illness detection, time, money wastage, and security; less cost; improved pharmaceutical management; and precise information to insurance providers regarding a client's health [1–5].

Globally, IoT in the healthcare market was estimated at about $98.4 billion USD in 2016 and is anticipated to grow at a compound annual growth rate of 27.3% over the estimated period of time. Some of the key factors, such as the aging population and

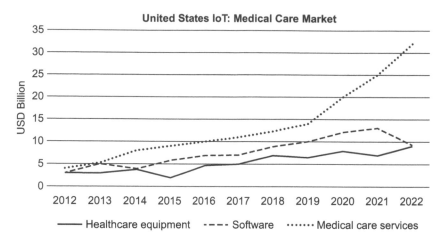

FIGURE 16.1
United States IoT: Medical care market by 2012–2022 (USD billion).

technological advancements, are likely to boost the IoT sector in the medical care market [4–6]. Growing awareness regarding improvements in disease-management tools such as telemedicine and remote monitoring of patients, as well as advancements in medical outcomes is also expected to boost the global market over the forecast period as shown in Figure 16.1.

Inpatient monitoring emerged as a major application segment in 2014. With increasing demands of disease management and incessant patient care services, effective treatment results are key factors contributing to the leading revenue share for real-time scenarios [6]. Telemedicine is another important segment in terms of revenue shares.

16.2 A Foresight on E-Healthcare Trailblazers

Over the last decade, while India's economy has accelerated at about 7%, the nation's healthcare sector continues to lag behind the technology. Shockingly, per the Indian Medical Association (IMA), the country's healthcare expenditures never surpassed 1.2% of total expenditures. This is even more worrying when compared with the US's 17% or our beloved neighbor China's 5.5%. Currently, India has only 0.7 doctors and 1.1 beds for every 1,000 of its citizens. The government aspires to increase healthcare spending to 3% by 2022. Much of this will be achieved through public-private partnerships and using technology generated by upcoming startups to increase the reach and multitude of e-healthcare solutions. A few examples of many are indicated below:

- **Achira Labs:** Achira Labs is a Bangalore-based startup whose primary focus is the development of a proprietary lab-on-chip platform to perform rapid, quantitative, and multiplexed immunoassays. Its innovative fabric chip, or "Fabchip," platform uses existing expertise and capital infrastructure available in the handloom textile industry to manufacture biological sensors using woven fabric. Achira's first product, ACIX-100, is a proprietary platform that has been calibrated to perform point-of-care tests for hCG (pregnancy), thyroid hormones, and female fertility hormones [7].

- **Bagmo:** Bagmo is a Bangalore-based startup working towards addressing issues of poor blood transfusion facilities in rural areas in India. Its first product is a novel blood bag monitoring solution to be used while blood is being transported to rural areas. Blood storage centers (BSCs) address the issue of blood availability to most of the remote locations in India. Bagmo is developing a networked smart temperature tracking system for blood bags. Each blood product (blood bag) comes with a unique radio frequency identification (RFID) and every BSC storage unit has an RFID scanner and efficient, smart temperature sensors. The proposed solution will ensure reliability of blood products [8].

- **Empathy Design Labs:** Empathy Design Labs is a Bangalore-based company working on developing a fetal monitoring device to prevent stillbirths in developing countries. As per WHO report, 2.6 million stillbirths take place every year across the globe. More than 98% of these stillbirths occur in developing countries such as India. Empathy Design Labs is designing a device to prevent stillbirths in developing countries by introducing an affordable, wearable, effective, and non-invasive screening and monitoring tool. After completing early clinical studies, the company plans to introduce the product into the Indian market in late 2019 or 2020 [9].

- **Jiva Sciences Private Limited:** Jiva Sciences is a startup working in the area of microfluidics and lab automation, currently developing solutions in the area of animal and human cell type identification. Its current focus is in the following areas: (1) microfluidics and photonic-based cell sorting, (2) microfluidics and photonic-based point-of-care diagnostics, and (3) lab automation. Jiva Sciences primary focus as an organization is in tune with the "Make in India" objective, through focusing on building hardware that can contribute towards issues of national importance as well as fit the technical dearth in India [10].

- **Mimyk:** Mimyk is a spin-off from the Indian Institute of Science (IISc), Bengaluru. The company works on developing immersive medical simulation technologies. It combines hardware and software to build next-generation simulation platforms. The current focus at Mimyk is an endoscopy simulation platform. Mimyk's technology is developed through a collaborative research and development activity carried out at the M2D2 Lab, IISc, and the Visualization and Graphics Lab (VGL), IISc [11].

- **Module Innovations:** Module Innovations is a Pune-based company developing strip-based biosensors for the detection of E. coli in food and clinical samples. Module aspires to bring healthcare parity by making systems that can detect diseases at the point of care, without needing a laboratory, trained personnel, or electricity, but with easy-to-interpret results with colorimetric readouts. Its product portfolio currently includes three devices: EcoSense (for E. coli detection), USense (detection of specific uropathogens), and CSense (screening for cervical cancer) [12].

- **Nesa Medtech:** Nesa Medtech is a medical device startup working towards excellence in the research and development of medical devices. It is developing an accessible, minimally invasive procedure to treat symptomatic uterine fibroids safely and effectively. Uterine fibroids are very common noncancerous growths that develop in the muscular wall of the uterus. Nesa proposes to develop a minimally invasive device for in-office use to treat symptomatic uterine fibroids. This can be achieved by ablating fibroids using a radio frequency electrode inserted through the uterine cavity under local anesthesia with real-time guidance through a transvaginal ultrasound. Nesa has secured follow-on funding from Government of Karnataka (GOK) through the Idea2 Point-of-care (POC) scheme [13].

- **Prantae:** Prantae is a Bhubaneshwar-based startup working on the development of a diagnostic platform for the miR-based detection of pre-eclampsia. Pre-eclampsia (PE) is one of top four causes of maternal deaths during pregnancy. Prantae proposes to translate the potential of identified miRs into an affordable clinical platform for early prediction of PE. It aims to develop a plasmonic biosensor detection optical unit capable of measuring a specially engineered gold nanoparticle-based biosensor harboring a DNA probe specific to PE miR. The proposed innovation is expected to have a significant socio-economic impact. Early diagnosis of PE not only ensures timely intervention during pregnancy but also will impact the expenditure on neonatal care due to PE-induced preterm emergency deliveries [14].

- **Shilps Sciences:** Shilps Sciences is a Bangalore-based startup focused on harnessing the power of nanotechnology instruments to solve problems in healthcare. It is developing an atomic force microscope and microfluidic technologies for isolating single cells for applications in life sciences. Shilps Sciences has developed a novel microfluidic technology to isolate and arrange cells. The company is now combining this with an atomic force microscope for cancer diagnostics. Shilps has created IP, built prototype demonstrators, and is in an advanced stage of product development. It has ongoing collaborations with cancer labs [15].

- **Unilumen Photonics:** Unilumen Photonics is a young venture currently incubated at IIT Madras that specializes in the manufacture of fiber laser systems. By partnering with key research institutions across India, it aims to develop solutions that address the biomedical and bioimaging needs of our society. The product portfolio includes pulsed and CW fiber lasers with low, medium, and high power optical fiber lasers. Unilumen's BM series biomedical laser product is capable of delivering high-power optical pulses suitable for photo-selective vaporization of prostate tissues, retinal photocoagulation, dental bleaching, treatment of facial vessels, and a broad range of other clinical applications. The company has commercialized its fiber laser products and has started generating revenue [16].

- **TerraBlue XT:** TerraBlue XT is a Bangalore-based company working on technologies to monitor mental health and manage epilepsy. TerraBlue XT's portfolio includes TJay, an award-winning device for the prediction, detection, and management of people with epilepsy. TJay catapulted the company to fame, when it became one of the top three winners of the Innovate for Digital India Challenge in 2015, supported by Intel and the government of India, which saw participation for 1900+ startups and individuals. TJay was quickly followed by Xaant, a wearable that helps practice calm, gives intriguing insights about the state of a person's mind, and shows how to beat stress. With its ability to understand mental health quickly, Xaant, currently in the final prototype stage, is expected to create waves in the wellness sector and is poised for release in the global market [17].

- **Yostra Labs:** Yostra Labs is a healthcare technology company mainly focused on diabetic care, with two products on the anvil. It is developing a market-viable diabetic peripheral neuropathy screening device called Sparsh that makes screening of diabetic patients more accessible and affordable to patients of all socio-economic strata. Yostra is also working on a device for the treatment of diabetic foot ulcers—Kadam, which enables faster healing of chronic ulcers [18].

- **Affigenix:** Affigenix is a Bangalore-based company with a focus on precision diagnostic kit development for personalized medicines. Its current portfolio includes a trypsin clearance assay kit. In addition, Affigenix is becoming a preferred service partner to a few biopharma companies and also earning a reputation for professionalism, performance, and productivity. It has also developed unique best-in-class critical reagents (22 #s) that are sensitive and specific to measure and monitor the safety and efficacy of products in preclinical, early, and late-stage clinical trials. It has commercialized >22 reagents and kits and has started generating revenue [19].

- **Aten Porus Lifesciences:** Aten Biotherapeutics has developed a degradable cyclodextrin polymer (DCP) platform of therapeutics that has shown to have therapeutic efficacy against Niemann-Pick type C (NPC) disorder in NPC patient-derived fibroblast cells. Because the DCP platform is based on nanoparticles, they are expected to have superior pharmokinetic profiles hence lowering the dosage drastically and improving patient compliance. Aten Porus has also started a company in the US [20].

- **Bugworks:** Bugworks aims to discover novel biopharmaceutical assets for the treatment of antibiotic-resistant bacterial infections. A spin-off from Cellworks Research, Bugworks uses a systems biology approach to discover novel treatment options for hospital-acquired infections. Through its innovative approaches, it has developed a stealth strategy by which antibiotics can successfully bypass efflux pumps, which are a key defense barrier present on the bacterial cell envelope. Its current leads target the WHO 2017 Critical, Serious, and Concerning infection threats. Bugworks is making progress on a first-in-class novel chemical entity (NCE) that exhibits potent killing of pan-resistant superbugs and difficult-to-treat pathogens [21].

- **Eyestem:** Eyestem is using stem cell and gene editing technologies to create breakthrough therapies for degenerative diseases of the eye. Age-related macular degeneration (AMD) affects the retinal pigment epithelium (RPE) layer of the retina and ranks third among the global causes of visual impairment. Eyestem's vision is to create scalable and affordable stem cell therapy for these degenerative eye diseases. It has successfully grown RPE and photoreceptors—two layers of the retina that are key for vision [22].

- **GeNext Genomics Private Limited:** Genext Genomics (GNG) is a Nagpur-based emerging life science research startup and contract research organization. GNG is working towards the development of recombinant proteins and monoclonal antibodies against clinically relevant targets through the integration of bioinformatics, recombinant DNA technology, proteomics, and immunology using novel high-throughput technologies. Its in-house research focus involves identifying and validating a drug target for tuberculosis; it aims to develop a mAb-based drug for TB. It has started generating revenue and has licensed one of its biopesticide formulations to Dharti Agro Fertilizers [23].

- **Nurture Earth:** Nurture Earth is an Aurangabad-based company that aims to serve as an Industry-academia bridge bringing in interdisciplinary technology development and engineering projects for execution by industry members and students/faculty in a collaborative manner. Nurture Earth works with international and local organizations as a contract and collaborative research partner. Its in-house research focus is on testing the feasibility of using nano-magnetic discs for ablation of cancer cells in vitro. This project was conceived as an application of nanotechnology in a nonsurgical targeted cancer cell removal procedure. Nurture Earth has recently started generating revenue through its collaborative projects [24].

- **Pentavalent Biosciences:** Pentavalent Biosciences is a resident startup mainly focusing on the development of novel vaccine molecules against fatal human diseases. Pentavalent is developing a vaccine formulation, comprising five major influenza strains including type A and type B, in replication-incompetent form, for intranasal administration in human populations. The resultant vaccine strains will be replication-incompetent, yet mimick wild influenza virus strains in structure. Due to this structural mimicking, the vaccine formulation comprising all five major strains will evoke a highly promising protective immune response in the host. The vaccine will target multiple populations in both seasonal and pandemic influenza [25].

- **Pandorum Technologies Private Limited:** Pandorum Technologies is a Bangalore-based company that develops proprietary science and technology platforms with the vision to manufacture personalized homo-chippiens and human organs on demand. Pandorum recently created artificial liver tissue by incorporating hepatocytes in biomaterial; this research has important implications in drug discovery. Pandorum's artificial three-dimensional liver tissue (organoid) efficiently recapitulates human liver tissue at the structural and functional level [26].

- **Theramyt Novobiologics Private Limited:** Theramyt's strategic goals include a steady progression on the value chain towards biobetters and new biological entities. Theramyt Novobiologics is developing two unique technology platforms for antibody engineering that are critically important to achieve the next generation of monoclonal antibody therapeutics. It has already filed six provisional patents related to its product and platform development technologies [27].

- **Viravecs:** Viravecs is a resident incubatee focusing on developing transgenic model systems for life science research. Currently, it is working on a novel technology to generate transgenic eukaryotic cell lines with no off-target effects [28].

- **Western Range Biopharmaceuticals:** Western Range Biopharmaceuticals is an Ahmedabad-based biotech startup developing treatment modalities for cancer patients. It has built proprietary technologies for cancer cellular immunotherapies, personalized chemo-sensitivity tests, GvHD reduction, and screening for HPV infection, one of the major causes for cancers in both men and women. Based on a blood sample, the tests will provide clinically relevant information [29].

- **Robust Herbals:** Robust Herbals is a Bangalore-based startup with a focus on developing Ayurveda's herbal tradition into a range of proprietary technologies and formulations as alternatives for existing pharmaceuticals/drugs. In addition, it is developing a biodegradable biocompatible blood pool contrast agent based on a rodlike nanoparticle constructed from FDA-approved materials with superior contrast enhancement [30].

- **Biomoneta:** Biomoneta is currently working on the design of a novel air purification device for the reduction of airborne bacteria. One of the ways healthcare-associated infections are spread in ICUs is via aerial transmission. Biomoneta is working on an air purification device that specifically reduces airborne bacterial contamination in ICUs and sick rooms. The device exploits the electrical properties of bacteria, novel antimicrobial fabrics, and engineering design; it is meant to be a bedside unit that both prevents an infectious patient from spreading the disease and protects a naïve patient from acquiring an infection [31].

16.3 Conclusion and Future Work

This chapter provides a look into the future of ehealth using IoT networks and also provides details about several ongoing projects that are working in this area of research with the goal of betterment of ecosystems. To date, countless manufacturing companies make devices for medical purposes for sharing medical data using sensory devices from patients to the nearest health centers. This sharing and analysis of medical-related data have revolutionized the healthcare organizations [32–55]. Medical practitioners will be able to provide timely guidance and provide prescriptions for their patients in a more convenient manner in case of emergency. In the future, we will work on sharing data over a network using Raspberry Pi and analyze the efficiency, accuracy, and robustness of shared data at the other side.

References

1. https://www.itu.int/en/ITU-T/academia/kaleidoscope/2014/Pages/SpecialSessionE-health,-IoT-and-Cloud.aspx
2. Sharma, L., Yadav, D., & Singh., A. (2016). Fisher's linear discriminant ratio based threshold for moving human detection in thermal video. *Infrared Physics & Technology*, *78*, 118–128.
3. Sharma, L., & Yadav, D. (2017). Histogram-based adaptive learning for background modelling: moving object detection in video surveillance. *International Journal of Telemedicine and Clinical Practices*, *2*(1), 74–92.
4. Sharma, L., & Lohan, N. (2019). Performance analysis of moving object detection using BGS techniques in visual surveillance. *International Journal of Spatio-Temporal Data Science*, *1*(1), 22–53.
5. Motion based object detection using background subtraction technique for smart video surveillance. (2018). http://hdl.handle.net/10603/204721
6. https://www.grandviewresearch.com/industry-analysis/internet-of-things-iot-healthcare-market
7. Microfluidics based point-of-care platform for immunodiagnostics. (2018). https://achiralabs.com/acix100/
8. Blood bag monitoring device. (2018). http://bagmo.in/about.php
9. A fetal monitoring device to prevent still births. (2018). https://www.empathydesignlabs.com/
10. Developing solutions in the area of animal and human cell type identification. (2018). http://jivasci.com/

11. Developing immersive medical simulation technologies. (2018). http://www.mimyk.com/
12. Strip-based biosensors for the detection of E. coli in food and clinical samples. (2018). http://www.moduleinnovations.com/?page_id=177
13. Symptomatic uterine fibroids treatment. (2018). http://nesamedtech.com/#aboutus
14. Diagnostic platform for the miR-based detection of pre-eclampsia. (2018). http://prantaesolutions.com/
15. Atomic force microscope and micro-fluidic technologies for isolating single cells for applications in life sciences. (2018). http://shilpsciences.com/
16. Biomedical and bioimaging. (2018). http://unilumen.in/bio-photonics/
17. To monitor mental health and manage epilepsy. (2018). http://www.teblux.com/tjay.html
18. Diabetic peripheral neuropathy screening device. (2018). http://www.yostra.com/
19. A trypsin clearance assay kit. (2018). http://www.affigenix.com/internalresearchfocus.php
20. Nanotechnology based therapeutics for lysosomal storage disorders and other rare diseases. (2018). http://www.atenporus.com/intellectual-property/
21. Treatment of antibiotic resistant bacterial infections. (2018). http://bugworksresearch.com/elude.html
22. Age-related macular degeneration (AMD). (2018). http://eyestem.com/science/
23. Development of recombinant proteins and monoclonal antibodies against clinically relevant targets through the integration of bioinformatics, recombinant DNA technology. (2018). http://gng.asia/gng/table.php
24. Nano-magnetic discs for ablation of cancer cells in vitro. (2018). http://nurtureearth.in/TechnologyDevelopment%20.aspx
25. Novel vaccine molecules against fatal human diseases. (2018). http://pentavalent.in/about_us.html
26. Technology platforms with the vision to manufacture personalized homo-chippiens and human organs on demand. (2018). http://pandorumtechnologies.in/science-tech/
27. Biologics and biobetters. (2018). http://theramyt.com/
28. Stable transgenic cell lines using CRISPR-Cas9 based genome editing and viral vectors. (2018). http://viravecs.com/
29. Treatment modalities for cancer patients. (2018). http://wrbio.com/?page_id=17
30. Ayurveda's herbal tradition into a range of proprietary technologies. (2018). http://www.robustherbals.com/
31. A novel air purification device for the reduction of airborne bacteria. (2018). http://biomoneta.com/
32. Riazul Islam, S. M., Kwak, D., Kabir, M. H., Hossain, M. S., & Kwak, K. S. (2015). The Internet of Things for health care: A comprehensive survey. *IEEE Access, 3*, 678–708.
33. The cellworks biosimulation platform. (2018). https://cellworks.life/technology
34. VAPCare: AI based secretions and oral hygiene management device. (2018). https://www.coeo.in/
35. A platform technology for healthcare solutions using 3D printed medical models. (2018). http://www.df3d.com/
36. Noxeno, which is a novel, dedicated, easy to use device for effective removal of impacted foreign bodies (NFB) from the nasal cavity. (2018). http://innaccel.com/
37. Soft tissue core biopsy (BioScoop™) and BxSeal™ technology. (2018). http://www.indiolabs.com/technologies.html
38. An imaging-based point-of-care diagnostic device. (2018). http://sminnovations.in/#portfolio
39. Diagnostic pacifier which has been developed as a platform for saliva based screening of various neonatal health ailments. (2018). http://www.spotsense.in/
40. Clinical genomics and hepatotoxicity prediction platforms. (2018). https://strandls.com/strand-clinical-diagnostics/
41. A platform for DNA synthetic vaccines. (2018). http://www.geniron.in/
42. Ren, J., Guo, Y., Liu, Q., & Zhang, Y. (2018). Distributed and efficient object detection in edge computing: Challenges and solutions. *IEEE Network, 1*, 1–7.

43. OGC Sensor. (2016). OGC sensor things API standard specification. http://www.opengeospatial. org/standards/sensorthings

44. Brown, E. (2016). Who needs the Internet of Things. https://www.linux.com/news/ who-needs-internet-things

45. Aribas, E., & Daglarli, E. (2017). Realtime object detection in IoT (Internet of Things) devices. In 25th Signal Processing and Communications Applications Conference (SIU), pp. 1–6.

46. Felzenszwalb, F., Girshick, R. B., McAllester, D., & Ramanan, D. (2010). Object detection with discriminatively trained part based models. *IEEE Transactions on Pattern Analysis and Machine Intelligence, 32*(9), 1627–1645.

47. Hu, L., & Giang, N. (2017). IoT-driven automated object detection algorithm for urban surveillance systems in smart cities. *IEEE Internet of Things Journal, 5*(2), 747–754.

48. Handte, M., Foell, S., Wagner, S., Kortuem, G., & Marrón, P. (2016). An Internet-of-Things enabled connected navigation system for urban bus rider. *IEEE Internet Things Journal, 3*(5), 735–744.

49. Li, B., Tian, B., Yao, Q., & Wang, K. (2012). A vehicle license plate recognition system based on analysis of maximally stable extremal regions. In Proceedings of 9th IEEE International Conference. Network Sensors Control, pp. 399–404.

50. Quack, T., Bay, H., & Gool, L. (2008). Object recognition for the Internet of Things. In IOT Lecture Notes in Computer Science book series LNCS 4952, pp. 230–246.

51. Right sensors for object tracking (2016). https://www.hcltech.com/blogs/right-sensors-object-tracking-iot-part-1

52. Detection and ranging for the Internet of Things. (2018). http://www.kritikalsolutions.com/ internet-of-things

53. Enabling detection and ranging for the Internet of Things and beyond. (2018). https:// leddartech.com/enabling-detection-and-ranging-for-the-internet-of-things-and-beyond/

54. Overview of Internet of Things. (2018). https://cloud.google.com/solutions/iot-overview

55. Internet of Things. (2017). https://arxiv.org/ftp/arxiv/papers/1708/1708.04560.pdf

17

Future of Internet of Things

Pradeep K. Garg[1] and Lavanya Sharma[2]

[1]*Civil Engineering Department, Indian Institute of Technology, Roorkee, India*

[2]*Amity Institute of Information Technology, Amity University, Uttar Pradesh, Noida, India*

CONTENTS

17.1 Introduction

The IoT consists of things, objects, devices, and machines that are connected to the Internet over fixed and wireless networks, which are able to collect data and share data with other devices. The IoT has the capability to connect infrastructure that was previously unconnected. As organizations and companies start using these connections to create value and provide services, it will be possible to add functionality to IoT networks to control the devices/infrastructure remotely [1]. In addition, IoT can automate the decision-making process within a defined range of input data. The IoT has the potential for redesigning business processes, and ultimately, reframing the working of business, government, and society. Businesses should start small, think big, and then scale fast in order to fully explore the potential of the IoT. By empowering businesses with the rapid scalability and flexibility essential to the IoT experience, growth opportunities abound in new and existing data.

Organizations having forward thinking are focusing on IoT initiatives and on developing approaches and new business models for managing the huge amounts of data, while leveraging IoT infrastructure. Many businesses are developing uses for IoT that directly impact humans or the environment, such as increasing food production, lowering carbon emissions, and enhancing the quality of health services. The development of new sensors can provide quality data in quantity, which is required to start new IoT-based applications [2]. Advantages of new sensors will be their tiny size with low cost, low power consumption, high reliability, and adaptability into IoT-based devices.

IoT has great potential in smart cities, transportation, health, education, buildings, environmental or energy efficiency, or combinations of these (Figure 17.1). In all such cases, geospatial technology can play a major enabling role. In addition, machine-to-machine (M2M) communication is a flexible technology that can use common equipment in newer ways to create an intuitive, cutting-edge, and intelligent system in place [3].

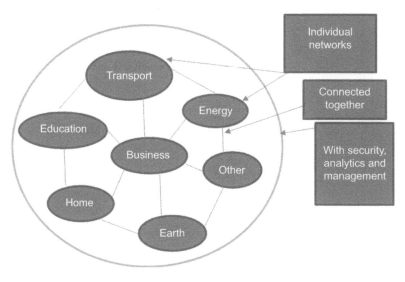

FIGURE 17.1
Potential of Internet of Things.

17.2 Growth of IoTs

IoT technology has drawn the attention of many network and communication researchers and scientists in recent years. It is estimated that twenty quintillion (10^{18}) bytes of data are being generated every day. Connected devices already outnumber the world's population by 1.5 to 1. The 25 billion machines today, by the year 2020 are expected to increase to 50 billion machines talking to each other, each with a unique IP address. In other words, not only people but things also will be interconnected. In 2004, G. Lawton [4] wrote, "There are many more machines in the world than the people, and a growing number of machines are networked... M2M is based on the idea that a machine has more value when it is networked and that the network becomes more valuable as more machines are connected."

Gartner, Inc. [5] mentioned that around 8.4 billion connected "things" will be in use worldwide in 2017, up 31% from 2016, and it is expected that 14.2 billion connected things will be in use in 2019, and that this number will reach 20.8 billion by 2020 and 25 billion by 2021, producing immense volumes of data. From 2018 onwards, multi-industry devices, such as those required at smart buildings, including LED lighting, heating, ventilation, and air-conditioning, and physical security systems, have taken a front seat, as connectivity generates higher volumes of data with lower costs of devices. In 2020, multi-industry devices are expected to reach 4.4 billion units, while vertical-specific devices will amount to 3.2 billion units. While consumers will purchase more devices, businesses will spend even more. By 2020, hardware purchases from both segments will reach almost $3 trillion.

In 2016, 5.5 million new things were connected to a network infrastructure each day. As IoT grows, the volumes of data it generates also grew. It is estimated that connected devices will generate 507.5 zettabytes (ZB) of data per year (42.3 ZB per month) by 2019, up from 134.5 ZB per year (11.2 ZB per month) in 2014. Globally, the data created by IoT devices in 2019 will be 269 times greater than the data being transmitted to data centers from end users' devices, and 49 times higher than the total traffic of data centers [6].

The continued use of M2M technology will boost the development of new business models. Research suggests that their applications are dramatically increasing and the number of devices connected would be around 40 billion. The economic impact and benefits of IoT are so huge that Gartner reckons it will exceed $1.9 billion by year 2020 [2, 5].

Evans [2] in year 2011 predicted that there will be about 28.4 billion IoT devices connected in year 2017, and this number is expected to almost double in next 3 years, as shown in Figure 17.2. It is projected to amount to 75.44 billion worldwide by year 2025.

17.3 Challenges and Issues of Concern

Connectivity to the Internet is becoming an increasingly common requirement in order to use IoT-based applications. However, Internet connectivity may not be necessarily available in some situations and locations, and still the power of the Internet is needed to reap the benefits of the IoT. The solution may be (Figure 17.3) that the data are transported from nodes that gather data from remote locations and deliver the data to a central hub that has Internet connectivity [7]. Hubs will collect all the data from different nodes and upload them online. The hub is therefore located in a major city with Internet access. The nodes

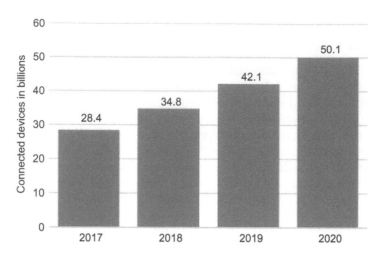

FIGURE 17.2
Number of connected devices worldwide.

can upload all the data to the Microsoft Azure Service Bus, from where it is easily accessible on a global scale. An upcoming technology in IoT, the Azure Service Bus is a great messaging service to transport messages and large data as well as store logs in the form of tables. The data from the Service Bus can then be accessed from anywhere in the world.

There are a range of scenarios of data transportation from nodes to hub, depending on the availability of transport resources: (1) A person could drive by road transport through the nodes to gather the data and drop them off at the hubs. Here, a simple USB drive/external SSD could be used in situations where some sort of human interaction is possible. (2) An unmanned aerial vehicle (UAV) or something similar could be used to transfer the data between nodes and hubs. The UAV would have a Wi-Fi card that would automatically

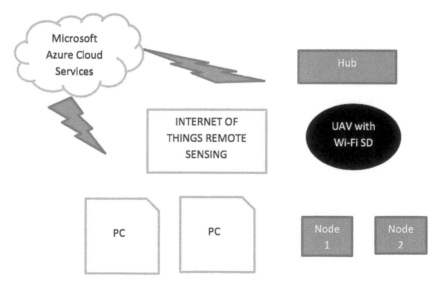

FIGURE 17.3
With limited or no Internet: Connectivity of data.

receive the data from each node, and transmit the data at the hub from which it is made available to citizens through cloud services. There can be endless possible implementations in the real world, from transporting medical records from nodes installed in hospitals to home automation and security as well as agriculture and greenhouses [7]. The innovation here lies in the automation involved with the integrated use of Microsoft's Azure Service Bus, where no or limited human interaction is required, e.g., plugging in the USB, or controlling the UAV [8].

IoT poses various challenging issues associated with communication nodes (things), large amounts of processing data, large numbers of communication nodes (things), routing, security, energy consumption for active nodes, and things coverage [9]. Each and every challenge has an important impact on the performance of IoT systems. These challenging issues may also affect the level of coverage that depends on communication technology between IoT nodes. The ground communication between things has a very restricted coverage range due to various impairments. Many existing technologies for IoT nodes have limited coverage (a few meters) [10]. The communication between IoT nodes using satellites or high-altitude platforms (HAPs) will provide much higher coverage, especially with Internet signals, which represent a main connection channel in IoT systems.

At present, the higher cost of IoT equipment, huge data storage requirements, real-time event processing using big data, and the use of geospatial databases pose a big challenge in IoT-based projects. Real-time analysis of geospatial big data currently can be done by software such as Microsoft StreamInsight [11] and ArcGIS GeoEvent Server [12]. Other research challenges include semantic web concepts to sensor discovery [13], which will turn the web into an information system to provide location-related information [14]. Another challenge for future IoT-based research is to develop methods that combine heterogeneous datasets to create four-dimensional models—three-dimensional models interfaced with time to enable users to view changes over time.

Some challenges while applying IoT-based techniques are discussed below [15–17].

17.3.1 Learning about Things

At present, we can learn about things through the Internet by reading and searching. As sensors become more widespread, we will be able to learn about things directly, maybe using our smartphones to interrogate things for information [2]. An example includes learning about a used car before we buy it, finding out its history, who owned it, how many kilometers it has run so far, what crashes it has had in the past, what parts have failed, how it has been driven over a number of years, and so on. At the moment, we can get some of this information via the Internet, but in the future it could be possible to discover all these details with a single push of a button.

17.3.2 Monitoring Things

Sensors can be used to monitor things. The IoT can combine a number of sensors and data from them to provide more knowledge and information [3]. A key potential area of use of sensors for monitoring is health or traffic. Smartphones could possibly be used to monitor our activity, our diet, our heart beats, and a number of other things. This could then be combined to give us regular updates on our health and even warn us when any health problem is expected.

17.3.3 Controlling Things

Smart meters are a good example of how things can be controlled through the IoT. For example, appliances in the house could be linked together and communicate to operate at the most efficient times for energy usage, while we are not present. On roads, traffic lights could be controlled using IoT in case there is an incident on the road [4].

17.3.4 Managing Things

Many tasks are being done in smart homes and smart cities, where energy and resources are managed in a more efficient way. More than half of the world's population now lives in urban areas, consuming a majority of the resources; therefore, there is a need to make our energy use more efficient [6]. In most cities, there is a serious problem with traffic congestion; if sensors and IoT devices are used to make traffic flows more efficient, this would have a great impact on resource savings.

17.3.5 Searching Things

Global positioning system (GPS) sensors provide locational data that can then track movement on smartphones or computers. If we can tap into the existing array of sensors already in place around the world, it may be possible to search and find almost anything.

17.3.6 Connectivity and Technical Issues

Internet coverage connectivity is continuously increasing, but still many parts of the world (particularly developing countries) are poorly covered with insufficient bandwidth. In those areas, IoT-based systems employing sensors are of limited use as they cannot transmit information in real-time. This is of great concern often in developing countries or remote areas. As IoT is expected to contain billions of devices around the globe, it demands ubiquitous network coverage globally, at all places. This can be obtained reliably through dense satellite networks [13, 14].

17.3.7 Security

Concerns have been raised worldwide about the security issues of the IoT. Without sufficient security testing of newly developed devices, hackers might be able to access and steal sensitive and personal data and to manipulate devices or parts of the network. This is even a more serious concern with the multitude of devices intended to be connected to the IoT, such as household devices, traffic, factories or hospitals, and medical sensors. Manufacturers of the components of IoT are required to design universal regulations to address this issue, which should include secure encryption for all involved devices [6].

Data security risks will vary with IoT, which cannot be overruled. Organizations and companies employing IoT will have to ensure that the information/solutions are secure, vigilant, and resilient. It is therefore necessary for them to understand what information is to be made public so that it does not have potential risks [18]. Every new device added in an IoT system will have an added opportunity for malicious attack, thus adding additional threats to devices, data, and users in the organization. Similarly, an unauthorized source trying to gain access to the system using the correct credentials may pose threat problems.

If any device fails to work or develops a bug, it should fail in a safe way without causing harm to the entire system.

Another potential drawback to this is viruses. Things interconnecting could be corrupted by viruses passed on to any number of things. This is probably the biggest challenge to the IoT, but at the same time it creates an opportunity for software developers to create fixes [17].

17.3.8 Confidentiality

Confidentiality is the main issue for the IoT system. In many applications, huge data are required to be generated from a variety of sensors, where IoT systems may access and analyze these data dynamically [16]. Proper management and the capability to handle classified data from specific devices can increase the chances of maintaining the confidentiality of the system. Procedures such as encryption can be applied to maintain confidentiality [13, 19].

17.3.9 Integrity

Frequent damages or failures of physical devices define the integrity of the system. Its protection is important, which may require precautions against sabotage. Data integrity will mainly depend on the robustness and fault tolerance of the IoT system. For example, in sensor networks, many RFIDs are not used all the time. This means they may be accessed by external attackers who can easily modify the data [18].

17.3.10 Availability

Availability and reliability of IoT systems are highly linked [20]. The IoT system should be able to show the performance levels in order to sustain availability. Software should be able to provide simultaneous services to users at any given location, whereas the presence of hardware should be there all the time for the application.

17.3.11 Interoperability

In order to for a number of machines and sensors to communicate with each other, they must understand each other, and must be compatible [16]. Many times, the variety of devices involved have different specifications. Therefore, universal and open standards, and protocols for interoperability within the IoT are needed for compatibility [21]. For this, the industries producing these devices will have to ensure interoperability, while keeping their systems proprietary.

17.3.12 Reliability

Reliability is essentially needed for deployment of IoT. Reliability of the system is important when Internet is available [22]. A network of up to billions of devices is highly complex, which also requires efficient management and regular upgrades. Examples include determination of a task when a device or a subnetwork discontinues functioning or becomes faulty. It may be caused by unstable electricity or fluctuations in power, which is another big challenge in developing countries affecting continuous Internet connectivity.

Dedicated networks with cognitive machine intelligence capabilities are needed to over-come this problem [23].

17.3.13 Speed

IoT applications may require exchange and integration of data between interconnected objects/devices so that faster decisions can be made. Therefore it demands high-speed data to provide fast results [23].

17.3.14 Low-Cost IoT

There is a need to develop low-cost communication propositions and sensors for large quantities of users for various applications [15].

17.3.15 Energy Demands

If more devices are linked to the IoT, more energy will be consumed. In addition, the storage of data generated by these connected devices requires sufficient energy. One promising application of the IoT therefore could be to monitor other devices for energy efficiency [24]. For example, sensors should not be kept permanently active if they are not being used continuously. Also, data can be stored for longer periods only when this is required for entire periods. Normally, most of the data generated within the IoT are needed only for a short time to send signals from M2M; storage may not be required.

17.3.16 Electronic Waste

Electronic waste is a serious issue worldwide, and the IoT is greatly contributing to it. Often, electronic devices have a short life span, and they need to be replaced when the device is not working or a device is available with new features. Electronic waste often contains hazardous elements, and recycling is not possible. These challenges affect developing countries in particular, because their Internet connectivity tends to be poorer, electricity supply unstable, and options to manage or recycle electronic waste are less developed. The problem is further aggravated by the high demand for trained skilled personnel to operate innovative new machines, and also for maintenance and repair of devices. Therefore, solutions for efficient energy use of the IoT and reduction of electronic waste are to be developed.

17.4 Future Application Areas

The IoT is still not fully matured. It is due to a number of factors, which limit the full use of the IoT. The future of the IoT seems to be boundless [17]. It includes the technology drivers and key application outcomes for the future [25]. A few initiatives in the area could play an important role in the success of this fast-emerging technology. Some of the key potential uses for IoT are given below.

17.4.1 Agriculture

Increasing demand for food puts tremendous pressure on agricultural production. Yields and productivity in agriculture can be increased by using IoT-based methods of increasing crop production. Also the latest technology can be used for its storage and distribution [26]. In developing countries, initial capital costs involved have a significant impact on the growth of agricultural productivity. M2M/IoT can very well overcome these restrictions. For current projected growth, satellite-based M2M/IoT terminals are being driven by developed countries, particularly North America; the Middle East and Africa forecast 5.2% annual growth. Stronger growth in agriculture in Asia can see revenues doubled by 2023, growing the region's share from 17.6% in 2013 to 23.2% in 2023 [27].

17.4.2 Construction Industry

In the construction industry, IoT-based system can furnish complete details and photos of the structure with time. It also helps in locating threats to construction caused by environmental degradation, earthquake, engineering works, landslides, mining, and industrial activities. IoT-based surveillance saves lot of physical visits to the actual construction site. This application in the construction industry is projected to grow 25% per year to $3.4 billion by 2022 [23].

17.4.3 Healthcare

In developing countries, the decreasing costs of sensors will soon enable the larger benefits of IoT. Reduced costs of sensors will enable the enhanced use of IoT technology in healthcare, such as in smart bandages. Here, built-in sensors in bandages would alert the patients and doctor if a wound or surgical incision is not healing properly. Very small aperture terminal (VSAT) networks also play an important role to provide backup when terrestrial networks fail, especially in critical applications in developing countries, such as healthcare.

17.4.4 Geospatial Data Mining

Some IoT applications need data from geospatial analysis. Geospatial mapping is gaining popularity not only to create maps of urban areas, traffic, pollution, and areas of interest [28], but also for disasters, such as earthquakes and landslides [29]. Urban emergency and risk maps could be prepared based on crowdsensing from users' mobile phones for assessment of risks [22]. In the future, real-time urban sensing could be done for personalized assistance while travelling [30] using data mining and geospatial tools.

In artificial intelligence (AI), extracting useful information at different spatial and temporal resolutions from a complex environment is a challenging research problem. Current methods use shallow learning methods with supervised and unsupervised learning [28]. In machine learning, research has to be focused in the field of deep learning [30], which aims to learn multiple layers of abstraction that are used to correctly interpret the given data. Furthermore, the resource constraints in sensor networks also create opportunities to use deep learning.

An important application area of IoT and geospatial data would be to assess the impact of climate change on urban areas [24]. Environment monitoring can be done, focusing on assessing and mitigating the impacts on human, land resources, livestock, and agriculture [27]. Transportation systems, route planning, traffic estimation [31], as well as safe driving [32] would be heavily dependent on the availability of better IoT infrastructures and powerful geospatial tools.

17.4.5 GIS-Based Visualization

Creative visualization can be created with the development of new display technologies. The evolution in display technology has given rise to highly efficient data representation using touch-screen interfaces where users are able to navigate the data better and more quickly than ever before. With emerging 3D displays, visualization of processes or event will certainly have more research opportunities. However, the data available from ubiquitous computing is not always ready for direct use for visualization platforms, and it may require further processing. It is very complex for heterogeneous spatio-temporal data [33]. New visualization schemes for the representation of data from heterogeneous sensors in a 3D landscape that also vary temporally are required to be developed [34]. Another challenge of visualizing the data collected within IoT is that they are geo-related and are sparsely distributed, and therefore a framework based on Internet GIS would be required to meet the challenges.

17.4.6 Satellites

Existing geostationary satellites provide terabytes of data worldwide, mainly used for direct-to-home broadcast and Internet over satellite connections. The effective use of geostationary satellites in IoT applications will employ large terminal antennas with enough gain to close the link and with sufficient directivity to avoid interference into adjacent satellites and systems. While the majority of IoT networks are terrestrial, IoT applications also provide various opportunities for connecting remote areas that lack terrestrial infrastructure as well as selling additional capacity on geostationary (GEO) satellites in C-, Ku-, and Ka-bands for direct or backhaul connectivity to deploying new low earth orbit (LEO) or highly elliptical orbit (HEO) satellites, optimized for the IoT market. The low latency of L-band services has a distinct benefit in catering to realistic applications, such as remote asset observing that requires reliable, always-on connectivity [1].

17.4.7 Sensor Fusion

Advances in sensor fusion lead to new applications, including smart homes and smart healthcare. However, these applications might generate significant privacy concerns, which must be understood for IoT governance. Sensor fusion platforms can be used to provide completely new services. For example, if sensors are installed on fruit and vegetable cardboard boxes, these could track the location, temperature, and jerks during movement and sniff the produce, and warn in advance of its spoilage.

IoT development requires that the environments, cities, buildings, vehicles, wearables, and portable devices have more and more information associated with them so that they are able to produce new information. The 5G scenarios reflect the future challenges and are expected to guide future endeavors in this area.

17.5 Conclusion

Data are like fuel that provide powers to the IoT system. An organization's ability to derive meaning from data will define its ultimate success. The IoT holds profound potential to provide data. In the future, the connected home, connected workplace, connected vehicles, and connected government will be a reality. The IoT has enormous scope for businesses. As IoT continues to develop, further potential is estimated by combining the related approaches, such as cloud computing, future Internet, big data, robotics, and semantic technologies to reveal greater synergy. In order to promote faster growth of IoT, key issues such as identification, reliability, connectivity, privacy and security, and interoperability must be considered. An open and integrated IoT environment is expected to increase competitiveness and make people's lives easier and more comfortable. For example, it will be easier for patients to receive continuous care. This will lead to better services, huge savings, and a smarter use of resources. To achieve promising results from IoT, the data protection legislation and the cyber security strategy need to be carefully planned and implemented. The highest growth potential in automation is expected in developing countries to work with the newer technology, which will also have great impact on applications. The IoT and AI are expected to play a vital role in the future. Organizations, industries, governments, engineers, scientists, and technologists have already started to implement both the technologies in many critical applications. The potential opportunities and benefits of both AI and IoT can be experienced when they are combined, both at the device end as well as at the server end. Data are only useful if they create an action, and to make data actionable, they need to be supplemented with IoT and AI together—*connected intelligence* and not just *connected devices*. As Internet usage spread worldwide in a short period of time, similarly it is expected that the IoT will make our lives better, with sensor fusion having a key role in the IoT system. By 2023, it is expected that new special-purpose chips will be developed that will reduce power consumption. Silicon chips enabling functions, such as embedded AI, will in turn enable organizations to create highly innovative products and services.

References

1. Said, O., & Masud, M. (2013). Towards internet of things: survey and future vision. *International Journal of Computer Networks, 5*(1), 1–17.
2. Evans, D. (2011, April). *The Internet of Things: How the next evolution of the internet is changing everything* (CISCO White Paper). https://www.cisco.com/web/about/ac79/docs/innov/IoT_IBSG_0411FINAL.pdf
3. Machado, H., & Shah, K. Internet of Things (IoT) impacts on supply chain APICS Houston Student Chapter. http://apicsterragrande.org/images/articles/Machado
4. Lawton, G. (2004). Machine-to-machine technology gears up for growth, *Computer, 37*(9), 12–15.
5. Gartner Inc. (2017, January). https://www.gartner.com/
6. Daecher, A., & Schmid, R. (2016, February 24). Internet of Things: From sensing to doing—think big, start small, scale fast. https://www2.deloitte.com/insights/us/en/focus/tech-trends/2016/internet-of-things-iot-applications-sensing-to-doing.html

7. Rajdev, U. (2014). Internet of Things remote sensors cloud collecting data from remote nodes with a WiFi-enabled UAV. https://udoo.hackster.io/umangrajdev/internet-of-things-remote-sensing-f2ce52

8. Rajdev, U. (2015). Internet of Things—Remote sensing in healthcare. https://www.hackster.io/umangrajdev/internet-of-things-in-heathcare-3ee264

9. Bari, N., Mani, G., & Berkovich, S. (2013). Internet of things as a methodological concept. In Proceedings of the 4th International Conference on Computing for Geospatial Research and Application (COM.Geo '13), San Jose, CA, pp. 48–55.

10. Ishaq, I., Moerman, I., Hoebeke, J., & Demeester, P. (2012). Internet of things virtual networks: Bringing network virtualization to resource-constrained devices. In Proceedings of the IEEE International Conference on Green Computing and Communications (GreenCom '12), Besancon, France, pp. 293–300.

11. https://msdn.microsoft.com/en-us/library/ee362541

12. https://enterprise.arcgis.com/en/geoevent/

13. METIS, Mobile and wireless communications enablers for the twenty (2020) information society. https://www.metis2020.com/

14. Kulkarni, R. V., Förster, A., & Venayagamoorthy, G. K. (2011). Computational intelligence in wireless sensor networks: a survey. *IEEE Communications Surveys & Tutorials*, 13, 68–96.

15. The Internet of Things: An overview, understanding the issues and challenges of a more connected world. (2015, October). www.internetsocuety.org

16. Furness, A. (2008). A framework model for the Internet of Things. GRIFS/CASAGRAS.

17. Internet of things in 2020: A roadmap for the future. 2016. https://www.smart-systemsintegration.org/public/documents/publications/Internet-of-Things_in_2020_EC-EPoSS_Workshop_Report_2008_v3.pdf

18. Mendez, D., Papapanagiotou, I., & Yang, B. (2017). Internet of Things: Survey on security and privacy. Cornell University Library.

19. Alam, S., Chowdhury, M. M., & Noll, J. (2011). Interoperability of security enabled Internet of Things. *Wireless Personal Communications*, 61(3), 567–586.

20. Roman, R., Zhou, J., & Lopez, J. (2013). On the features and challenges of security and privacy in distributed internet of things. *Computer Networks*, 57(10), 2266–2279.

21. Brunkard, P. (2018). The future of IOT is AI. (2018, September 3). https://www.techuk.org/insights/opinions/item/13827-the-future-of-iot-is-ai

22. Liu, J., Shen, H., & Zhang, X. (2016). A survey of mobile crowdsensing techniques: A critical component for the internet of things. In 25th International Conference on Computer Communication and Networks (ICCCN).

23. Satellite technologies for IoT applications. (March 2017). Report produced by IoT UK.

24. Gubbi, J., Buyya, R., & Marusic, S. (2013). Internet of Things (IoT): A vision, architectural elements, and future directions. *Future Generation Computer Systems*, 29(7), 1645–1660.

25. Atzori, L., Iera, A., & Morabito, G. (2010). The Internet of Things: A survey. *Computer Networks*, 54, 2787–2805.

26. Lakshmi, K., & Gayathri, S. (2017). Implementation of IoT with image processing in plant growth monitoring system. *Journal of Scientific and Innovative Research*, 6(2), 80–83.

27. Kamilaris, A., Assumpcio, A., Blasi, A. B., Torrella, M. & Prenafeta-Boldu, F. X. (2017). *Estimating the environmental impact of agriculture by means of geospatial and big data analysis: The case of Catalonia*. Springer International Publishing.

28. Parks, D., Mankowski, T., Zangooei, S., Porter, M. S., Armanini, D. G., Baird, D. J., Langille, M. G. I., & Beiko, R. G. (2013). GenGIS 2: Geospatial analysis of traditional and genetic biodiversity, with new gradient algorithms and an extensible plugin framework. *PloS one*, 8(7).

29. Zook, M., Shelton, T., Graham, M., & Gorman, S. (2010). Volunteered geographic information and crowdsourcing disaster relief: a case study of the Haitian earthquake. *World Medical and Health Policy*, 2(2), 7–33.

30. Bengio, Y. (2009). *Learning deep architectures for AI*. Now Publishers Inc.

31. Richly, K., Teusner, R., Immer, A., & Windheuser, F. (2015, November). Optimizing routes of public transportation systems by analyzing the data of taxi rides. In Proceedings of the 1st International ACM SIGSPATIAL Workshop on Smart Cities and Urban Analytics, Seattle, WA, pp. 70–76.
32. Lee, U., Zhou, B., Gerla, M., Magistretti, E., Bellavista, P., & Corradi, A. (2006). Smart mobs for urban monitoring with a vehicular sensor network. *IEEE Wireless Communications, 13*(5).
33. Bonneau, G. P., & Nielson, G. M. (2003). *Data visualization: The state of the art*. London: Kluwer Academic.
34. Ren, L., Tian, F., & Zhang, X. (2010). A model-based user interface toolkit for interactive information visualization systems. *Journal of Visual Languages and Computing, 21*, 209–229.

Index